# Student Projects in Environmental Science

# Student Projects in Environmental Science

**Stuart Harrad and Lesley Batty**

*Division of Environmental Health and Risk Management,
School of Geography, Earth, and Environmental Sciences,
University of Birmingham, UK*

**George Arhonditsis and Miriam Diamond**

*Department of Physical and Environmental Sciences, Department of
Geography, University of Toronto, Canada*

John Wiley & Sons, Ltd

*Other Wiley Editorial Offices*

John Wiley & Sons Inc., 111 River Street, Hoboken, NJ 07030, USA

Jossey-Bass, 989 Market Street, San Francisco, CA 94103-1741, USA

Wiley-VCH Verlag GmbH, Boschstr. 12, D-69469 Weinheim, Germany

John Wiley & Sons Australia Ltd, 33 Park Road, Milton, Queensland 4064, Australia

John Wiley & Sons (Asia) Pte Ltd, 2 Clementi Loop #02-01, Jin Xing Distripark, Singapore 129809

John Wiley & Sons Canada Ltd, 6045 Freemont Blvd, Mississauga, Ontario, L5R 4J3

Wiley also publishes its books in a variety of electronic formats. Some content that appears in print may not be
available in electronic books.

*Library of Congress Cataloging-in-Publication Data*

Student Projects in Environmental Science: Conducting your research project in quantitative
environmental science / Stuart Harrad ... [et al.].
       p. cm.
   Includes bibliographical references and index.
   ISBN 978-0-470-84564-6 (acid-free paper) – ISBN 978-0-470-84566-0
(acid-free paper)
1. Environmental sciences–Research. 2. Quantitative research.  I.
Harrad, Stuart, 1962–
   GE70.C63 2007
   628–dc22                                         2007047556

*British Library Cataloguing in Publication Data*

A catalogue record for this book is available from the British Library

ISBN 9780470845646 (HB)
ISBN 9780470845660 (PB)

Typeset in 10.5/12.5 pt. Times by Thomson Digital, Noida, India

# Contents

# Preface

This book arose from the authors' perception of a need for a text that would provide simple, clear guidance to students on how to conduct their research projects in the quantitative environmental sciences. While designed primarily for use by undergraduates, it is also likely to be of assistance to those studying for an MSc, and to those in the early stages of their PhD.

The book provides guidance on the key stages involved in conducting a research project. Beginning with a discussion of what constitutes research, it shows how to develop and plan a feasible research topic based upon a testable hypothesis. It then stresses the importance of logical experimental design that facilitates the use of appropriate statistical treatment of data so that the research questions posed may be answered.

To reinforce key points, use is made throughout of illustrative case studies covering a wide range of topics. It is important to note that the book does not include detailed (and often confusing) theoretical treatments of the commonly employed statistical tests (students are referred elsewhere to suitable texts where necessary). Instead, it develops student understanding of what statistical tests are appropriate to use when analysing their data, and offers practical guidance on how these may be conducted using widely used commercially available software such as Excel, Minitab, and SPSS.

Given the increasing importance of mathematical modelling in the environmental sciences, an entire chapter is devoted to helping students to understand the fundamental principles involved in model development, testing, and interpretation. Care is taken to demystify these processes and the jargon associated with this area.

Finally, guidance is given on how to write a dissertation, as well as a range of transferable research skills, including presentation skills, how to conduct oneself in an oral examination, and how to "manage" one's supervisor!

In conclusion, it is the authors' intention that this book should provide a concise yet comprehensive guide for students undertaking research in the quantitative environmental sciences. We hope that you, the reader, will find it so.

# Acknowledgement

The authors gratefully acknowledge Mr Alan Whitfield of the Faculty of Education, Health, and Sciences at the University of Derby, UK for contributing the illustrative case-study 5.1.2.

# 1

# General strategies for completing your research project successfully

## 1.1   Introduction – why is this book necessary?

This is a legitimate question. Effectively, after all, this is a 'textbook' on how to conduct quantitative research in environmental science. If we accept that there is a need for quantitative research in environmental science – one need only make reference to issues such as reduced biodiversity, habitat destruction, global warming, stratospheric ozone depletion, and chemical pollution of our air, water, and food to win that argument – then we are left asking why teach research methods? Certainly, it is only relatively recently that serious thought has been given on a widespread basis as to how research methods can and even should be taught in anything like a systematic fashion. At the time that the authors conducted their BSc, MSc, and PhD research (the 1980s), there was no systematic programme teaching research methods. Instead, the authors (and there is no reason to believe that their experience was in any way exceptional) were expected to pick up how to conduct research by osmosis. Although diehards could and do suggest that such a macho 'sink or swim' approach must have had some merits, given the success of some of those subjected to it, it is surely more likely that the success of such individuals occurred in spite of the absence of formal research training, rather than because of it.

Thankfully, a more enlightened approach has evolved in recent years, and in recognition of the importance of research skills, the tuition of research techniques is now an increasingly widespread component of the curriculum of undergraduate degree programmes. Furthermore, the desirability of research training and its provision of transferable skills, is now fully acknowledged by the UK Research Councils, who require that universities in receipt of Research Council studentships must have in place a programme of research training for all PhD students, and who now also sponsor Master of Research degree programmes that emphasise the tuition of research techniques.

*Student Projects in Environmental Science*   Edited by Stuart Harrad and Lesley Batty
© 2008 John Wiley & Sons Ltd

Embarking on your first (or second or even third, depending on your academic history) research project can be envisaged as a 'rite of passage', in which you progress towards becoming a mature scientist. By mature, we mean an individual capable of independent and creative work. The creative aspect of research – which is often overlooked by practitioners of the natural sciences due to a misconception that creativity is something limited to the humanities – cannot be overstated. After all, one can work independently while learning about what is known and not known about the causes of global warming, but it is only when one expands the pool of knowledge on such a topic, by generating a new idea (and thus being creative), that one can truly claim to be a researcher. One of the primary purposes of teaching research methods (and thus of this book) must therefore be to clarify what is meant by research, and to differentiate it from the (largely dependent and certainly not creative) scholarly processes of learning that you will have previously utilised. Making such a transition from scholar to researcher requires the acquisition of a range of distinctive skills that most students with little or no research experience will not have previously encountered. Becoming a researcher can be likened to joining a Club. This Club – the research community – has a complex system of rules governing its functioning. Hence, by explaining the rules, rather than leaving them to be discovered serendipitously, the teaching of research methods can speed up the process by which you become an active and effective member of that Club.

This book aims to provide as much assistance as is possible to enable you to complete successfully your dissertation. However, at various points throughout this text we recommend that you consult more specialist texts or help facilities for guidance on specific matters such as how to use Excel, Minitab, or SPSS software packages, and detailed explanations of statistical techniques.

## 1.2   What on earth am I going to do for my research project?

In this respect, PhD students in environmental science may be considered lucky, as they are assigned a research project. At undergraduate and Masters level this is not usually the case and students will have to play a large part in devising a suitable topic for their research project. The key to coming up with a good project lies to a large extent in its viability, as well as its scientific merit. To illustrate, while a research project aimed at proving or disproving a causal link between air pollution and exacerbation of respiratory illnesses such as asthma is of undoubted scientific merit and concern; it is wholly inappropriate for a lone undergraduate to undertake, given the modest time period, resources, and research experience available to them. It is thus vital that you seek advice from a potential supervisor at the earliest possible stage. They will be able to advise you as to the feasibility and scientific validity of your ideas. The feasibility of an idea as an undergraduate/Masters research project cannot be overlooked. You will need to consider a variety of aspects: the availability when needed of data and/or equipment required to produce data; the costs involved (budgets are rarely large!); the amount of time required to complete the project relative to that available; and your skills (both practical and theoretical) compared with those necessary to conduct the research.

Similarly, if you are stumped completely for ideas, supervisors will usually have some potential topics for you to consider and discuss. Overall though, the importance of seeking and taking advice from those with experience cannot be overemphasised. In particular, doing so at the earliest possible stage can prevent you wasting valuable time. One suggestion is that if you can, consult students who have already undertaken or are undertaking similar projects, as such a 'student's eye view' may help.

Table 1.2.1 gives several examples of research ideas that – although valid scientifically – are not viable as an undergraduate project because of the excessive amount of time, resources and/or expertise required for their completion. For each such idea, a corresponding 'undergraduate-friendly' and viable version is provided. Essentially, these are more highly focused and compact versions of the 'bigger' idea.

In conclusion, it is worth spending time to develop a suitable project idea, and you are strongly recommended to take advice at the earliest possible opportunity.

**Table 1.2.1**  Examples of viable and non-viable objectives/strategies for undergraduate projects

| Non-viable objective/strategy | Viable objective/strategy |
|---|---|
| Monitoring campaign to examine seasonal and time trends of a parameter (e.g. rainfall, population of a species, pollution) over a number of years | Analysis of previously acquired and *available* datasets on seasonal and time trends covering a number of years |
| Assessment of typical UK adult exposure to one or more pollutants | Assessment of exposure of a small number of individuals (e.g. friends, family, fellow students) |
| Study to determine the human health impacts of exposure to small airborne particles (e.g. $PM_{10}$)[a] | Measurement of the level of exposure to $PM_{10}$ experienced by a small number of individuals, and comparison with relevant air quality standards |
| A fieldwork-based study of the impact of a major flood event | A desktop-based study of the impact of one or more major flood events, based on archived data obtained during a past flood |
| A fieldwork-based study of coastal erosion | A desktop-based study of coastal erosion over a number of years, using archived map records |
| Glacial hydrology | A study of suspended sediment dynamics in a specific glacier-fed stream |
| Investigation of links between water quality in UK canals and catchment land-use | Study of the relationship between water quality (as indicated by a few easily measurable parameters) and land use in a small stretch of one canal |
| Climate change | A critical review of the relationship between climate change and the frequency of extreme weather events in the UK, or a critical review of the possible change in disease patterns due to climate change in Europe |

[a] $PM_{10}$ refers to *airborne particles with an aerodynamic diameter less than 10 μm.*

## 1.3    Fundamentals of scientific research, the generation and testing of hypotheses (see also Chapter 4)

The generation and testing of hypotheses is the cornerstone of quantitative research in environmental science. In essence, it involves asking a question, suggesting a possible answer (or answers), and devising a strategy for evaluating (testing) the veracity of that answer or answers. In this book, we will refer to this 'suggest and test' approach, as the generation and testing of hypotheses. For example, previous research may have shown that airborne concentrations of a chemical pollutant are higher in winter than in summer. Our question would then be why? There are a number of possible (and plausible) reasons why this may be so: (a) emissions of the pollutant are greater in winter; (b) the depth of the boundary layer (i.e. the layer of the atmosphere closest to the surface) is shallower in winter than in summer; and (c) the greater solar irradiance in summer results in the pollutant being more readily degraded via photolysis or reaction with the photolytically generated hydroxyl radical than in winter. Any single one (or combination) of these reasons may explain the observation.

Our hypotheses would thus be that winter concentrations are higher:

(a) because emissions are greater in winter (this may be because the pollutant is generated by burning fuel to heat homes/offices, etc.);

(b) because the depth of the atmospheric boundary layer above the Earth's surface is shallower in winter – thus the volume of air in which pollutants emissions are diluted is less, and hence concentrations are increased;

(c) because photochemical losses (i.e. those due to reaction with sunlight) are less significant in the shorter and less sunny days of winter.

To test these hypotheses, we would need to evaluate the extent to which changes in concentration are attributable to changes in one or more of the following: emissions, boundary layer depth, and solar irradiance (sunlight intensity). This may be conducted relatively easily if data on concentrations, boundary layer depth, and solar irradiance are already available. However, if simultaneous measurements of these parameters have not been made previously, then it will be necessary to set up a monitoring campaign to measure them.

Related to these hypotheses is the question of whether or not concentrations of the pollutant really are higher in winter than in summer. In short, are concentrations statistically significantly higher? If previous data do not prove unequivocally this to be the case, then you will have to conduct a monitoring campaign to verify whether or not this does occur. Section 2.2 will help you design your campaign, and interpret the data it generates, in a way that will enable you to decide whether winter concentrations do exceed those in summer.

An important concept here is that of the **null hypothesis**. This centres around the assumption that the hypothesis that you are testing is incorrect. In the example in the paragraph above, our null hypothesis would be that concentrations of the pollutants are NOT statistically significantly higher in winter than in summer (you can perhaps think of this as equating to the concept of 'innocent until proved guilty'). Hence, if our results disprove the null hypothesis, then there is a statistically significant increase in pollutant concentrations in winter compared with summer. We will return to this concept in section 2.2.

## 1.4   What constitutes research? Distinguishing between monitoring and research

This is a difficult question to answer. At its simplest level, one could define research as furthering human knowledge. At this point, one should distinguish between monitoring or intelligence gathering, and genuinely enhancing understanding. To illustrate with reference to the example above, establishing that airborne concentrations of a chemical pollutant are higher in winter than in summer, constitutes monitoring or intelligence gathering. By comparison, interpretation of this monitoring information to elucidate **why** this is so, requires genuine understanding, and it is this latter activity which can be classified truly as research. To illustrate further, the chapter of your dissertation that contains your results would – on its own – provide useful data, but would not by this definition constitute research. It is only by interpreting these data that you move into the realms of research by evaluating your hypothesis and thus advance understanding. The term intelligence gathering was used above in the same context as monitoring. Both monitoring and intelligence gathering may be defined as the gathering of the information required to test a hypothesis. The distinction between the two terms is that monitoring refers to the collection of new information, while intelligence gathering refers to the collation of existing data. To refer back to our example, if previous monitoring had generated all the data necessary for us to test our hypotheses as to why winter concentrations exceeded those in summer, then our method of data collection would be intelligence gathering. Alternatively, if some or all of the data required to test the hypotheses were unavailable, then our method of data collection would be a monitoring campaign to generate the necessary data. The major difference between the monitoring approach and the intelligence-gathering approach to data collection is that the former is usually far more expensive and time consuming. It is largely for this reason that while many final year undergraduate and Masters level projects will be centred around monitoring, many involve intelligence gathering. By comparison, PhD projects are far more likely to involve monitoring. The great advantage of monitoring is that with careful design, **you** can ensure that all the information required to test your hypothesis is collected, rather than rely on the necessary information being available. This need not be an obstacle to intelligence-gathering, however, but you must make sure before you start your project that **all** the data you need will be available to you.

## 1.5    Project planning

While recognising the inherently unpredictable nature of research – by definition you are investigating the unknown – it should be feasible to plan and timetable the entire course of an undergraduate project. Of course, before finalising any plan, you should discuss its content with your supervisor, who will be well qualified to advise on how realistic it is.

### 1.5.1    Why plan?

There are numerous benefits from planning. These include:

- Making you think carefully about what it is that you are going to do, and in particular about the possible problems that you may encounter and how you may overcome them.

- Setting deadlines for the completion of key stages. This self-imposed discipline is of great value in ensuring that you avoid the all-too-common problem of running short or even out of time. This problem is the most common cause of students obtaining poor marks, or even no marks whatsoever. Given the likely major contribution of your project marks to your overall degree classification, you need clearly to do everything possible to maximise your grade. The comparatively long period between commencing their research project and the deadline for submission of their dissertation deadlines is something that many students have difficulty coming to terms with. The problem is that by definition, those who need help the most in this respect – i.e. those who have the greatest problems meeting deadlines – are those least likely to be motivated by the self-imposed deadlines of a research plan. This is why many degree programmes now include a number of externally imposed deadlines where your progress will be evaluated. If you feel that you are likely to ignore the self-imposed deadlines of a research plan, then you may wish to give your supervisor a copy of your plan, as – owing to the many other demands on their time – they are likely to be less flexible than you about meeting target dates.

- Provision of a yardstick against which you can compare your progress. In reality, it is quite possible that you will not adhere rigidly to your plan. At several points, you may find yourself either behind or ahead of schedule. When this happens, you should consider adjusting your plan to account for the changed circumstances.

### 1.5.2    How to plan

The following represents a checklist of points you should consider when planning your research:

- Identify goal(s) and consider whether they are realistic (see section 1.2).

- How much will you have to do? Will you need help in making measurements? For example does making the measurements involve the use of laboratory techniques with which you have no experience? If so, make sure that the necessary help is available before you start.

- Seek and take advice.

- Pinpoint possible problems and consider how they can be overcome.

- Identify 'milestones' of achievement required for completion of goal(s).

- Set realistic deadlines for the completion of each 'milestone'.

To illustrate how to devise a research plan, we will examine a case study based on a research project supervised recently by one of the authors (see Box 1.5.2.1).

---

## *Box 1.5.2.1*  Research-plan case study

### Background

While the course within which this 1-year project was conducted set aside formally a minimum of 6 weeks in which students were expected to devote themselves to collecting data, this period extended frequently to 12 weeks, with more time required for data analysis and writing the dissertation. It is also worth noting that during the last 4 months of the project, the student had to attend lectures and complete coursework. Depending on your personal loading in this respect, you may wish to allow more (or less) time than outlined here for those stages falling within this period.

### Project

The student had an interest in air pollution, and wished to compare concentrations of fine particulate matter ($PM_{10}$ – i.e. airborne particles with an aerodynamic diameter less than 10 $\mu$m) in their home town with the relevant air quality standard – at the time of writing 50 $\mu$g m$^{-3}$ as a 24-h average. In short, the project had two objectives: (1) to measure concentrations of $PM_{10}$ in a small provincial city, and (2) to identify sources of the $PM_{10}$ monitored and if possible rank them in order of importance.

Our first consideration should be to evaluate whether these objectives are feasible. In this respect, one has to bear in mind the time available for monitoring, the student's technical background (i.e. will there be sufficient time for them to learn any necessary specialist techniques?), and resource implications (e.g. costs and equipment availability).

After some initial enquiries, the student was able to borrow an instrument capable of monitoring $PM_{10}$ continuously over a 6 week period compatible with the student's availability. As the instrument was automated, and relatively simple to operate, it was concluded that objective (1) was readily achievable.

Identifying and ranking sources of pollution is commonly referred to as source apportionment. Given the importance of this activity (how can one hope to reduce pollution, if one does not know where it is coming from?), it is unsurprising that there exists a substantial array of source apportionment methods. Our task therefore would be to identify an appropriate source apportionment technique (i.e. effective but feasible given the time, student skills, and resource constraints of this project).

One technique commonly used is to compare the chemical composition of $PM_{10}$ with that of particles emitted from different sources. To illustrate, in the good old bad old days of leaded petrol (to which a bromine-containing compound was added), high concentrations of lead and bromine, were a clear indication that traffic emissions were an appreciable source of the particulate matter. Although – owing to removal of lead from petrol – it is no longer as simple as this, and other pollutants emitted by traffic need to be used as 'indicators' of the source of $PM_{10}$; it is, with careful statistical analysis, still possible to quantify the relative contribution of traffic emissions to $PM_{10}$. Unfortunately, in the context of this comparatively short project, the detailed chemical analyses required for implementing this technique would not be feasible. This is due both to the additional time and costs required for such analyses, and the time needed for the student to learn the techniques required to conduct the analyses.

An alternative technique relies on evaluating how concentrations of $PM_{10}$ vary with time of day and a variety of meteorological parameters. To illustrate, if concentrations follow a daily pattern displaying a peak during both morning and evening rush hours, then an appreciable contribution from traffic is implied. Similarly, if higher concentrations are observed when the wind comes from a certain direction, then the presence of a source or source(s) in that location is/are implicated. Hence, provided that $PM_{10}$ could be measured with adequate frequency, and meteorological data were available for the monitoring site, this latter option would be the way forward.

The key points that would be considered in the planning process of this project are summarised below.

1. Is necessary equipment available? Specifically:

   • Is a $PM_{10}$ measuring device available for the period required?

- Are meteorological data available?

2. Consider what the likely problems are. Specifically:

   - How reliable is the measuring device? What if it breaks down? Is it easily and quickly repaired, or is a spare readily available?

3. How will you measure $PM_{10}$?

   - Select a suitable sampling location (ground-level but secure, and with access to an appropriate power supply)

   - How frequently will you monitor and for how long? (ideally continuously if your equipment is capable of this, and for 6 weeks)

4. How will you identify sources?

   - Monitor concentration variations with wind direction and rainfall

Taking all the above into consideration, the student devised the following 'milestone-based' plan (see Table 1.5.2.1).

**Table 1.5.2.1**  Research plan devised for a project determining levels and sources of $PM_{10}$ (based on the project involving 6 weeks fieldwork commencing 22 weeks before the submission deadline)

| Milestone number | Target date[a] | Description |
| --- | --- | --- |
| 1 | 1 | Site sampler |
| 2 | 1 | Ensure meteorological data will be available |
| 3 | 7 | Finish monitoring |
| 4 | 11 | Finish plotting $PM_{10}$ against wind direction, speed, rainfall, etc. |
| 5 | 18 | Finish initial draft of report |
| 6 | 20 | Finish final draft |
| 7 | 22 | Submit dissertation |

[a]Beginning of week number.

## 1.5.3   How to implement your plan

Once you have drawn up your research plan, it is important that you utilise it. In summary:

- Follow your research plan, it is there to help you succeed.

- Review progress regularly, a suggested approach is to spend at least 15 minutes considering how things are going at the same time every week – e.g. Friday afternoon.

- Feed data back into the plan, this helps you adapt to developments as they occur, and to follow up interesting aspects that were not originally envisaged.

- Revise plan if you fall behind/surge ahead of schedule, it would not be unusual for your initial schedule to be either overoptimistic or pessimistic, so revise in the light of experience.

## 1.6   Conducting your project safely

Your safety and that of others during the conduct of your research project is hopefully self-evidently important. The supervisory procedures applied at your university should – at the very least – require you to think through with the assistance of your supervisor or other member of staff, any potential hazards involved in your project and how you will prevent or minimise the risk of you or others falling victim to those hazards. However, it is considered helpful to consider briefly such issues here.

The hazards and risks associated with some types of projects are more obvious than others. Those involving the handling of toxic or harmful chemicals or organisms in laboratories will require strict compliance with established procedures laid down by the laboratory within which you are working. It is essential that you make sure that you are familiar with these, that you understand them (do not hesitate to ask if there is anything you are unsure about in this respect), and that you are equipped with the necessary protective clothing (safety glasses, gloves, and laboratory coat). Likewise, for those involving fieldwork either in the UK or overseas, to acquire environmental samples will require strict adherence to established procedures. Furthermore, it is important to remember that even seemingly 'safe' projects can involve risk. For example, personal safety must be considered even when gathering data via questionnaires or interviews. The risks are more obvious if one is collecting data by hand, and one should never undertake such 'field' visits alone, and always let people know where you are going and the time you anticipate returning. Even if you are collecting your data via post, telephone, or even email, do not reveal your home address in any communications with respondents – e.g. use the university's address for return of questionnaires; even if you wish to analyse responses over a vacation period, you can always arrange for returned questionnaires to be forwarded to you at your home address.

## 1.7   How to conduct a literature review (see also Chapter 7)

One of the first steps that should be undertaken when deciding upon a viable topic for your research project is conducting a critical review of the scientific literature on the subject. This serves the extremely useful purpose of allowing the student to familiarise themselves with their subject area, and identify any gaps in knowledge or conflicts of evidence that their research project can try and fill or resolve. Furthermore, if performed well, it can form the basis of the first chapter of a dissertation or thesis. As with writing and research in general (see Chapter 7), one of the hardest aspects of conducting a literature review is getting started, given that there is likely to be a vast number of articles available. It is best to start with a seminal article/review on your subject, which your supervisor should be able to provide. Quite often, this 'starting point' will be the thesis or dissertation of one of your supervisor's previous students at either undergraduate or post-graduate level. Follow up references in this and/or use the Science Citation Index (SCI) to locate later papers that have referenced the original articles. The SCI is especially good when the original article is relatively old, as it helps locate reports of more recent developments in the subject area. From this solitary starting point, the quantity of research papers and reports you will uncover will mushroom rapidly. Sometimes the mountain of material of potential interest will expand to alarming proportions. If so, then it is best to step back and refocus on the reason for conducting your literature review at this stage – i.e. to decide whether there is a viable topic for you to research (see section 1.2). Frequently, reading all available literature is simply not feasible, and you are better advised to make use of good review papers of the topic – you will get no credit for reinventing the wheel. Again, this is an area where if you have any doubts as to what best to do, then you should ask for help from your supervisor.

Given the likely wealth of material out there, it is important that the review be conducted within an appropriate time constraint – i.e. compatible with finalising your agreed topic and supervisor before the deadline imposed by your specific course. A vitally important thing for the student to remember is that the review is a **critical** one. Its purpose is **not** to track down and report slavishly the findings of every paper ever written on the topic. Rather, it should consist of a concise critical review and summary of the current state of knowledge in the field. Box 1.7.1 gives a checklist of key points to critical evaluation of research papers.

---

## Box 1.7.1    Critical appraisal of research papers

**How scientifically objective is it? Are its conclusions well supported? To what extent does it consider all the evidence available and possible interpretations of that evidence? Are the authors basing their conclusions on the evidence, or are they interpreting their results solely in a manner that supports their hypothesis?**

For inexperienced (and even experienced) researchers this can be very difficult to assess, so it is best to seek advice from your supervisor where you are unsure.

Suspicions that this is the case may be raised by – amongst other things – where the authors work and/or the source of funding for the research, usually indicated in the Acknowledgements section. For example, one might consider it unusual for someone working for or funded by a tobacco company to publish research high-lighting the adverse human health effects of smoking. Similarly, research funded by an anti-smoking group would be unusual if it concluded that the health impacts of the weight gains experienced by some ex-smokers outweighed the benefits of quitting smoking. It must be stressed though that there is plenty of good research conducted by scientists within industrial or pressure group environments, and likewise that there is plenty of lower quality research carried out by workers within independent research institutions such as universities.

## In which journal was it published?

Look at its impact factor relative to others in the same research area. Impact factors are a method of expressing how highly researchers rate papers published in a given journal – the higher they are, the 'better' the journal. They are based – amongst other things – on the number of times that other papers cite a given paper, the idea being that a highly cited paper originates from a high quality journal. Generally speaking, a journal that selects papers for publication based on a peer-review process is likely to have greater merit than those that do not.

## The peer review process

A background knowledge of how research papers are evaluated as to their suitability for publication is useful in evaluating their quality. Scientific journals such as *Environmental Science and Technology* use a system known as 'peer review'. Funda-mentally, this involves submission of the paper to the journal's editor, who distributes copies to two or three scientists expert in the field covered by the paper. These scientists then review the paper for scientific validity and acceptance of the paper by the editor is based on the comments of these reviewers. Three outcomes may follow submission of a paper: (1) it may be rejected outright; (2) it may be accepted unaltered or (3) it may be accepted, provided that a number of specified changes (which may involve conducting additional work) are made. Out-comes (1) and (3) are by far the most prevalent. This process ensures that the technical content of the paper is assessed by people well qualified to do so and helps maintain the quality of published papers. Such a process is rarely implemented by 'lesser' journals or trade magazines, with the result that major errors are more likely to appear in articles published in such journals.

An important consideration is that only relevant material should be included. What constitutes **relevant** material is of course open to interpretation. However, it can be broadly defined as anything that either concurs with or contradicts the hypothesis to be tested. You may find several papers that say broadly the same thing. In such instances, your aim should be to summarise what it is that these papers are saying collectively, and report it in a style akin to this:

'Several studies (*inter alia a*, *b*, and *c*) suggest that their observed winter time peaks in concentrations of *x* are attributable largely to seasonal variations in atmospheric boundary layer depth. Conversely, other authors (notably *d*, *e*, and *f*) argue that the winter increases in airborne concentrations of *x* are predominantly due to enhanced emissions from combustion activities in winter'.

While the authors of these papers may be horrified to see their hard work summarised in so peremptory a fashion, such a summary is sufficient for your purpose. You may feel that the only way to demonstrate that you have read a wide range of papers is to report each in detail. This is a mistake, it not only makes the literature review unbearably turgid, but indicates that you have failed to understand one of the key purposes of the review; namely, to **critically** review and **summarise** the **relevant** literature. In the above example, the relevance of the literature reviewed is that it demonstrates that there exist conflicting opinions as to what causes wintertime peaks in concentrations of *x*. In so doing, it provides the ideal platform for justifying your research, by pointing out that you aim to resolve this conflict of opinions and generate the definitive answer.

## 1.8 How to be a research student

The following sections will discuss the challenges of and possible approaches to becoming a successful research student.

### 1.8.1 Making the adjustment from scholar to researcher

Until now, you will have attended lectures, tutorials, etc., and been assessed by sitting exams, and writing essays, etc. In almost every case, you will have been set a question, or given similarly clear instructions as to what to do. It is in this key respect that your research project differs from anything else that you have done so far. By its very nature, conducting research involves a degree of uncertainty. As you have progressed through each year of your degree programme, you will have increasingly become aware that simple 'black and white' answers to questions, are rarely satisfactory. To illustrate, with respect to the example of a research project given in Box 1.5.2.1, it will not be sufficient for you to simply collect data (either by monitoring or intelligence gathering) on airborne concentrations of a pollutant. You will have to interpret your data so that you are able to say something about: (1) whether there are seasonal differences – supported by some statistical analysis; and (2) why such differences (if any) exist or do not exist.

## 1.8.2   Making the most of your time – strategies for time management

Time management is very much a personal issue – like techniques for writing your dissertation (see section 7.1), there are many different yet equally viable approaches. However, there are some key aspects common to each approach. The first of these – and probably the most important – is making sufficient time. Before you can set about this task, you need to have a clear idea of how much time is required, and at what stage of the project. It is recommended that you get the advice of your supervisor on this issue at the earliest possible stage, as the exact requirements will vary from project to project and course to course. However, as an indicative guide, we include here the advice given by one of the authors to undergraduate students conducting projects requiring both the taking of samples in the field, and their subsequent chemical analysis in his laboratory. The time-frame within which such projects are conducted is between the end of the students' seccond year (mid-June) and the submission of the dissertation in the middle of the following February. Students are expected to spend 2–3 weeks taking samples, and a further 2–3 weeks conducting the chemical analyses. This practical work is conducted over the summer, usually (although not always) in two separate chunks (e.g. sampling in June/July, analysis in September). In this way, students should have the data needed to test their hypothesis or hypotheses by the beginning of their final year. The analysis and interpretation of these data and the writing of the dissertation is then conducted between October and mid-February, with students recommended to spend one full day a week every week on this.

If we use the above as an example, making time for the practical work over the summer holiday is comparatively simple, requiring only that the necessary time periods are kept clear of other work and holidays. A rather more difficult proposition is making time once the final year is in full swing, when it is necessary to balance work on your dissertation with deadlines for submission of other coursework, attendance at lectures, seminars, etc., part-time employment, and a social life. If possible, you should aim to spend a few minutes at the beginning of every week devising a written plan of how you will spend each morning, afternoon and evening. This approach allows flexibility from week to week and makes it easier to see when you can make time for various tasks. If you have a full-day slot free, then this is probably ideal for conducting your dissertation work, but clearly this is not always possible.

Table 1.8.2.1 gives an example workplan for a student in the first week of December of their final year (week 10 of the first semester).

**Table 1.8.2.1** An illustrative workplan

| Day | Morning | Afternoon | Evening |
|---|---|---|---|
| Monday | Lectures | Practical class | Coursework |
| Tuesday | Coursework | Lectures | Coursework |
| Wednesday | Lectures/tutorial | Sport | Leisure |
| Thursday | Lectures | Dissertation | Leisure |
| Friday | Lectures | Dissertation | Leisure |
| Saturday | Work | Work | Leisure |
| Sunday | Work | Work | Coursework |

### 1.8.3 What to do when the going gets tough

It is not unfair to state that 'research is not all glamour'. Indeed, it rarely is anything approaching glamorous – although it can of course, be extremely rewarding. However, it would be naïve to expect that your research project will progress smoothly to a successful conclusion without the slightest hitch, or that you will not feel bored, daunted by the magnitude of what remains to be done, or in some other way disillusioned with your project at some point. In addition to your partner, friends, and family, it is often well worth talking about the way you are feeling to both your supervisor, and other students, both those at the same stage, and also those who have passed the stage you are currently at. If you are conducting your research project within a large group of researchers, then postdoctoral researchers and PhD students are also a good source of practical advice and sympathetic ears, as they will all have been through something akin to what you are experiencing.

## 1.9 How to manage your supervisor

### 1.9.1 Why you need a supervisor

The short answer here is that having one will dramatically increase your chances of success. One of the most common reasons why undergraduate research projects fail to reach their full potential is that there has been insufficient supervisory input. While this could – and occasionally does – occur through no fault of the student, the student must take the blame in most instances. By this, we mean that the student has done one or any number of a variety of things, including:

- failed to keep appointments with their supervisor;

- failed to respond to requests from their supervisor to arrange such meetings;

- failed to take heed of advice from their supervisor, such as suggested changes to their project plan, methods, interpretation of results, and the writing and presentation of their dissertation.

The combination of sound advice and a good student who takes advice from their supervisor, will inevitably lead to a high quality dissertation for which the student will receive a good mark and of which they can be proud. Clearly, although the student's academic ability and hard work is the single most important factor in determining the success of a research project, the quality of the supervisory input plays an important role. The next section discusses what you can do to get the best supervision possible.

### 1.9.2    What your supervisor does when they are not supervising you

Contrary to widely held public belief, the typical academic involved in research is an extremely busy individual. If you as a student are to get the best out of your (very important) relationship with your supervisor, it is useful to have an appreciation of the other demands on their time.

The job title of 'lecturer' is arguably outdated, and is a prime cause of misunderstanding regarding the typical academic's job. Although some academics do spend the majority of their time involved in teaching, for many others this aspect of their work is a minor (but important) component. One thing that many undergraduate students undertaking their research project come to realise quickly is that academics do not 'have the summer off'. This is because there are postgraduate research students and postdoctoral researchers to supervise, research papers and applications for funding for new research projects to write, postgraduate teaching to be carried out, 'resit' examinations to set, invigilate and mark, new course material to write, old course material to review and update, etc., etc.

Instead, for many academics, a more appropriate job title might be 'research manager'. Certainly, at a given point in time the authors of this book each typically supervise 5–10 PhD students and postdoctoral researchers, and anything up to 10 undergraduate and taught MSc research project students.

Given the array of other demands on your supervisor's time, you might be forgiven for taking the view that your supervisor cannot possibly find the time to be interested in you or your project. Thankfully, this is rarely the case. To understand why this so, remember first that the project will be one that the supervisor will have played a part in designing (and so by definition have an interest in), and second, a successful outcome for a project will form the best possible basis for obtaining funding for further research and/or writing a research paper. Aside of these more base and (from the supervisor's angle) selfish reasons for taking a keen interest in your research, bear in mind the perhaps strange concept that your supervisor might actually like you. All in all, supervisors want their students to succeed, and a wise student will make use of that knowledge.

### 1.9.3    Getting the best out of meetings with your supervisor

A common experience of supervisors is for a student to launch into a description of a particular problem they are experiencing (or may even have solved) without any description of the background to the issue. In such instances, the student has overlooked that while they have (rightly) been focusing on (worrying about) one specific problem, their supervisor has been dealing with anything up to 20 similar topics. Given this, you should take pity and give your supervisor a chance to 'tune in' to your wavelength. One way of tackling this problem is for you to present your supervisor with a short written report in advance of your meeting, or at very least an indication of what you wish to discuss. Clearly, this approach is not always feasible, but in such instances, a short introductory sentence or two on entering the room along the lines of 'You remember that I was having problems with $x$ because of $y$, and you suggested that I try $z$, well, I did $z$ and this happened' before unleashing the full sad tale, will go a long way to helping your supervisor to offer you the prompt and sound advice that you are after. In summary, you need to give your supervisor time to think through your problem if they are to offer you the best possible advice. The best way of doing this is to arrange a meeting in advance. Once you have done so, make sure you turn up on time. Not doing so, is not only bad manners, but will not likely endear you to your supervisor and in many cases, will either mean that the length of your meeting (and thus the extent and quality of the advice that you receive) will be shortened, or that they will not be able to see you.

## 1.10   Summary

Like building a house, completing a successful research project requires you to lay firm foundations. Before commencing your project you need to devise a suitable project that is compatible with the resources available to you. You need to plan carefully how you will carry out the project and write up your findings within the time constraints of your course, and any part-time or vacation work that you have. Finally, you need to make use of your plan, review your progress regularly, and make good use of your supervisor. Following the advice in this chapter will represent a significant first step towards conducting a rewarding and successful research project.

# 2

# Gathering your data

## 2.1 Different types of data

Once you have decided on your general research area and defined the hypotheses you want to test, it may be tempting to launch straight into collecting the data. However, this will inevitably lead to poor datasets that are unsuitable for testing the hypotheses and cannot be analysed statistically. Therefore it is important to spend some time planning the research in detail (see also sections 1.5 and 2.2).

The first thing that you need to determine is the type of data required to address the aims and objectives of the project. Although in most instances it is much better to collect your own data, referred to as **primary data**, it may not be achievable within the time-frame or resources of the project. In this case you may need to use **secondary data**, which are data collected by other people. For example, you may want to determine whether there has been a significant change in nitrate concentrations within lowland rivers over the last 10 years. Obviously you cannot collect these data yourself and therefore you will need to use data that have been collected as part of ongoing monitoring; in the UK this type of data would be collected by the Environment Agency. These data will have originated from sampling and analysis by a number of different people and is therefore potentially subject to a whole range of problems associated with quality assurance and control (see section 2.3). However, it does allow long-term patterns to be identified outside the time-frame of a normal research project. It is essential when designing and planning this type of research project that one ensures that the relevant datasets are available and sufficiently complete for the requirements of the project (see section 1.5).

Whether one is using primary or secondary data, it important to understand whether it is **quantitative** or **qualitative**. In both physical and life sciences it is usual to collect quantitative data, which means simply that what you are measuring can be quantified in some way (it can be assigned a number). Qualitative data are used more often in disciplines such as social sciences, but may crop up in environmental

*Student Projects in Environmental Science*   Edited by Stuart Harrad and Lesley Batty
© 2008 John Wiley & Sons Ltd

sciences occasionally. For example, the use of kick sampling for macroinvertebrate analysis in rivers is only semi-quantitative because although a count of organisms is obtained, the volume of sediment that they come from (and thus the population density) is unknown. Truly qualitative data is simply descriptive data (e.g. information about the views of local residents about noise generated by a nearby airport).

Data can be classified further as **discrete** or **continuous** and there are several types of data within these divisions. It is essential to know which type of data you are planning to collect as this will determine the type of statistical analysis that can be used, influencing your outcomes.

## 2.1.1  Discrete variables

*Nominal*
In this case data can be assigned to a specific quantitative category. For example, soil type is an example of nominal data, although we use words in this case (laterite, podsol, brown earth, gley, etc.) these could easily be replaced by numbers (e.g. laterite = 1, podsol = 2, etc.). The importance of this type of data is that it allows comparison of data collected by many different people. However, when interpreting this type of data one needs to be very careful. For example, let us say we have assigned a numerical value to soil types as listed above and have examined the occurrence of these soils in a particular area. Once the data have been collected it may be tempting to analyse it numerically, for example working out the average value. However, this would mean absolutely nothing. Say we calculated the average value to be 2.6, this does not give the average soil type in the area. Thus it is important to be careful when working with nominal data. Other examples of nominal data could be eye colour or blood type.

*Binary*
This is a special type of discrete variable where data can be assigned to one of two states. For example, if one was measuring the occurrence of a particular species in a variety of habitats; the data would be assigned to either present or absent. Another example may be if you wanted to code males and females within a population, male could be 0 and female 1.

*Ordinal*
This is similar to nominal but in this case the categories can be ranked in some way. For example, we may classify river water quality as very bad, bad, poor, fair, good, very good or excellent, or air pollution as low, moderate, high or very high. Again this allows some sort of comparison between datasets but similar limitations apply as for nominal data in terms of mathematical analyses. Examples of ordinal data can often be found in questionnaire-based research, for example questions that require the respondent to indicate whether they disagree strongly, disagree, neither agree or disagree, agree or agree strongly, generate ordinal data.

## 2.1.2  Continuous variables

These are numerical data and therefore allow the use of mathematics to analyse the data further. There are two types of data here which are divided according to whether there is an absolute zero (e.g. mass, weight, or concentration) or alternatively, an arbitrary zero (e.g. temperature in degrees centigrade). Those data with an absolute zero are referred to as **ratio** data and the zero in this case means an absence of the thing that you are measuring. In contrast, data which have an arbitrary zero are referred to as **interval** data, and the zero in this case simply indicates a low value. When comparing interval data it is important that they are on the same scale, as a zero in one is not always equivalent to a zero on another scale. For example $0°$ on the centigrade scale is not the same as $0°$ on the Fahrenheit scale.

# 2.2  Designing an experimental research project

In terms of relative importance, the design of a project is arguably the most important part of research, because if you get this wrong then any statistical analysis and interpretation will either be impossible or nonsense! Therefore, the more time spent on this stage, the more likely it is that you will get valid results.

The thing to remember with all research is that you are trying to identify patterns and processes from a very complicated world. This means that there are lots of factors out there to influence the component that you are measuring and thereby complicate interpretation. Good experimental design will eliminate many of these factors but one can never get rid of them all and you must be aware of these when interpreting data. The second thing to remember is that in an ideal world we would be able to measure every single component of the system we are testing. In reality this is both absolutely impossible and undesirable, first because we simply do not have time and also because it would destroy the very system we are studying. Thus whenever research is carried out we are taking a **sample** from the entire population and we are assuming that the results from this sample reflect accurately the results that would be obtained if we analysed every single individual or component of the population. In other words, we need a representative sample. When we talk about **populations** we are referring to the entire information of a system from which we take a representative portion or sample. This is carried out in many surveys, for example when monitoring viewing figures for television programmes, information on viewing is gathered for a number of households that are representative of the whole population (it would be impossible to monitor every single household in a country). In the context of research and statistics a population cannot only be individuals (e.g. people or plants or insects) but also characteristics, events or anything else that gives rise to numbers. The use of samples to represent the entire population may seem to be a major assumption but there are many ways in which researchers can ensure that this is reasonable, all of which should be considered in order to create a good experimental design.

In order to test our hypotheses there are a number of different approaches that can be used, the choice of which will depend very much on the questions being posed. The first option is to collect data from 'real situations', the second is to carry out experiments under field or 'real life' situations, and the third is to carry out experiments under controlled conditions. So let us take an example to illustrate how these different approaches may be applied.

We may be interested in the effect of acidity upon the growth of algae in rivers and hypothesise that algal growth is limited in conditions of low pH (<4). The first terms to introduce here are those of the dependent and independent variable. A variable is essentially something that can vary, and the **independent variable** is the component of the experiment that we control or manipulate and the **dependent variable** is the thing that varies as a consequence of (or depends on) changes in the independent variable. In this case we are controlling the pH in some way, hence pH is the independent variable whereas we are measuring variations in algal growth in response to changes in pH and therefore this algal growth is the dependent variable (see also section 5.1).

So if we take the first approach of collecting data from 'real situations' then we would need to identify a number of rivers displaying different pH. Once these have been identified then we can measure algal growth within these environments by taking representative samples. There are a number of problems with this approach. First we need to have some knowledge of the environment in order to be able to select appropriate sites that cover a range of pH values (it may not be possible to locate sufficient suitable sites perhaps because we do not have enough time or money for the necessary travel), and second there are many other factors that will also vary between the sites apart from pH, such as nutrient concentrations, underlying geology, elevation, degree of shading, etc., which may also influence growth of algae. It is possible to minimise these factors through careful site selection but they cannot be eliminated altogether.

The second approach is to carry out experiments in 'real situations'. This would involve the use of sites as identified in the first approach (covering a range of pH) but instead of simply measuring what is already there we would measure the response in something introduced into the environment by us and over which we therefore have some control. For example, we may introduce blank tiles into the rivers and measure the growth of algae on these over time. The advantage of this approach is that we are exerting some sort of control over the system ensuring that there is some consistency across the different sites, i.e. the use of a blank tile means that the substrate for algal growth is the same at each site.

The final approach of using laboratory experiments allows us to maximise the control we have over the system and reduces the effects of confounding factors. What we can do here is to set up our own artificial river systems and alter the pH artificially ensuring that we have all the conditions that we want to test. We can also make sure that all other aspects of the systems are the same (such as temperature, nutrient concentration, etc.) so that if we detect differences in the growth of algae between the different systems we can ensure that it is due to pH and not another factor. In laboratory systems we can usually make much more accurate measurements than we can under field conditions, simply due to practicalities. However, laboratory experiments also have their drawbacks. The fact that they are artificial means that the results obtained often do not hold true in real environments and there is often a problem of scale. For example, work undertaken on weathering rates of rocks within laboratories has shown that rates can be order(s) of magnitude lower in the laboratory than for equivalent rock types in the field. This is due largely to the heterogeneity of environments found under field conditions that cannot be reproduced accurately in the laboratory.

Each of these approaches has advantages and disadvantages but the one chosen to test the hypothesis will depend very much on the details of the objectives of the study, how much previous data has been collected, as well as practicalities and resources available. Whichever approach is chosen, however, there are a number of considerations that need to be thought through in the planning stage (see section 1.5). The first thing to consider is the nature of sampling. It is simple enough to state that samples will be taken, but exactly how do we select which samples to take? Imagine if we wanted to determine whether plant growth changed in response to metal concentrations in soils. We could have a number of sites that we want to take samples from, but we need to know which samples (or plants) to measure. Now we could simply go to the site and choose plants but it is very easy to introduce **bias** in this case as we may unconsciously choose those plants that are larger and more obvious. Alternatively you may choose to sample plants in one small area, but what if metals are not evenly distributed throughout the site or that area has been affected by a herbivore? Again this has the potential to introduce bias to your sampling and make the conclusions you draw from your data highly questionable. The best way of carrying out sampling is by using **random** samples, where each sample is given an equal chance of selection. This is usually achieved through the use of random number tables. So in the example above, we would take a plan of the site to be sampled, divide it into small sections that are numbered and then use random number tables to identify which sections should be sampled. Of course in practice, this may not always work, for example in some sections there may not be any plants growing or you may not be able to access parts of the site. If this is the case, minor adjustments in the sampling regime may need to be made. This is perfectly acceptable but must be taken into account when analysing data and drawing conclusions.

In some cases it may not be appropriate to sample the entire population. For example if we wanted to establish whether plant growth varied between species growing on contaminated and uncontaminated soils we would need to subdivide the population into different species. This is because growth would vary between the different plant species due to natural genetic variation and thus it would be inappropriate to take all plants growing on the site as a single population. This is referred to as **stratified sampling**, where the overall population is divided into sub-populations and then random samples are taken from within each sub-population.

A similar approach may also be used where the entire population is divided into **clusters** and a random sample of the clusters is taken and all observations within each cluster taken. This approach may be used where taking individuals within an entire population which is so widely scattered that taking random samples would be too costly or time consuming. For example, if we wanted to examine growth of a particular plant species within the UK, it may occur at various locations that are widely scattered. In this case it would be better to identify individual populations or clusters of the species and then select randomly clusters to sample. A word of caution here; clustering of individuals may be related to a factor other than that you are measuring, which has the potential to yield erroneous results.

The second consideration in good experimental or sampling design relates to natural variation. Remember that we are trying to determine whether our result is real and not due to chance. Now let us go back to our example of plant growth on metal contaminated sites. What happens if we take one plant from each site and measure the biomass (dry weight) of the plant? Would this give us enough information about the growth of all plants within that site? The answer of course is no. Within a population there is natural variation (not all plants growing on the same site will be the same size) and measuring one plant will not be sufficient to quantify this natural variation. Remember we need to establish whether the variation within each population is less than that between the different populations (in this case the different sites). In order to quantify variation we need to take replicate samples, that is we need to take a number of samples from each environment in order to assess natural variation. The number of samples required will be determined from the power analysis as discussed later in this section, but may also be limited by logistics and resources. The same thing applies to laboratory experiments as to field data collection. If we were setting up artificial systems to test algal growth and had only one system for each pH condition, we would again have no idea whether the variation between systems subjected to different pH is greater than natural variation in growth that would be observed between systems experiencing identical pH. In this case we would require replicate systems.

When designing experiments one thing we need to ensure we avoid is **pseudo-replication**. This occurs if replicate samples or experiments are not completely independent. For example, if we had designed an experiment to determine whether copper concentrations in soil affected plant growth, then if we simply grew five individual plants in the same pot and then treated each plant as a replicate, this would be pseudoreplication as the plants are not independent of each other (they can actually influence the growth of the other plants). Instead what we would require is five individual plants in separate pots containing the same concentration of copper in order to provide five replicates. This is absolutely essential for statistics as one of the major assumptions of the majority of statistical tests is that samples are independent.

The use of replication allows us to ensure that natural variation is accounted for, but we only have variation in organisms that are subjected to changes in the variable we are testing. What we do not know is whether there is some other factor that is changing the response of organisms apart from the one we are varying. In order to make sure that this is not the case we need to include a **control population** in our experimental design. The control population or group is a set of subjects that do not receive the experimental treatment but in all other respects are treated identically (e.g. same temperature, dissolved oxygen content, nutrient levels, etc.). This is very easy to do under laboratory conditions but is much more difficult to achieve in field situations. So in the example used previously where we wanted to determine whether copper concentration affected plant growth we would need to include a set of pots that had not been treated with copper but in every other respect were grown in exactly the same conditions as those which had copper added.

Within any population there is natural variation (e.g. not all cats are the same size, not all plants of the same species have the same number of leaves, etc.) and therefore when we are taking samples from a population (e.g. of cats, a given plant species) we need to make sure that we have collected enough to characterise natural variation. If we do not do this then the 'background' variation in our samples will be too high for us to determine whether there are any relationships or differences in populations (see section 4.2). This leads us to a crucial aspect of experimental design. How do we determine the number of samples that are needed to characterise the population sufficiently? This is where it is important to carry out preliminary investigations. By doing this we can get an idea of the extent of natural variation within a population and by carrying out a **power analysis** we can determine exactly how many samples would need to be taken. So let us look at this in more detail.

When we design an experiment we are trying to ensure that we make the correct decision as to whether we reject or accept the null hypothesis (see section 1.3). There are two major types of error that we want to avoid when designing experiments, type I and type II errors illustrated in Table 2.2.1.

Type I errors are related to probability and the level at which we accept that something is statistically significant (see section 3.2), which is often set at 5%. Type II errors are related to the design of the experiment. What this means is that the sample mean of your data may actually be very different from the true population mean and as a result you may reject incorrectly the null hypothesis (see section 1.3). When designing an experiment it is important to assess how likely this is to occur by using power analysis.

The 'power' of the test (i.e. its ability to tell whether two or more populations are different – e.g. plant growth on one site is greater than on others) depends upon a number of factors including the variation within samples, the sample size and the type of statistical test that is applied. There is also something known as the 'effect size' which is a way by which the size of the difference between two groups is quantified

**Table 2.2.1**  Relationship between the null hypothesis in real and experimental situations

|  | | Null hypothesis is: | |
| --- | --- | --- | --- |
|  | | Accepted | Rejected |
| Null hypothesis in reality: | Accepted | Correct decision | Type I error |
|  | Rejected | Type II error | Correct decision |

(e.g. how much greater is the plant growth on one site compared with others), and essentially the larger this figure is, the greater the power of the test. There is little that one can do practically to alter the effect size but it is important to measure it accurately. In contrast, the type of statistical test and the sample size can both be manipulated in the experimental design to maximise the power of the test. If we can obtain estimates of the effect size and sample variation then we can use a simple test that is available on most statistical packages, to determine the sample size that is required to accurately detect the effect that we are trying to achieve. By doing so, we not only ensure that we collect enough samples for statistical analysis, but we also prevent wasted effort in collecting many more samples than are actually needed. It is also possible to carry out a power analysis retrospectively in order to determine the validity of any statistical results that have been generated. However, as such a retrospective test may bring bad news that you have not collected sufficient data, it is far better to perform it prospectively – i.e. before you start!

So to illustrate this process by example, a researcher wants to establish whether there is a difference in cadmium concentration in two different soils from sites A and B, in order to determine whether historical application of herbicides on one of the sites has affected soil quality. The first thing that is required is to make preliminary analyses to determine the variation of cadmium concentration within the population (in this case the population is the soil at each site). The data for five samples from sites A and B are presented in Table 2.2.2. See section 3.1 for details on how to measure variation in data using standard deviation and standard error.

The standard deviation in this case is found to be 0.018 for soil A and 0.84 for soil B. The next thing to determine is the **effect size**. This is essentially how much of a difference does the researcher want to establish between the two populations (this is referred to as **differences** in Minitab). This can be decided either by basing it on effect sizes other researchers have used for similar experiments (although this number is often not reported), on previous knowledge, or on preliminary experiments. If you use preliminary experiments to determine effect size then the equation required is:

$$\text{effect size} = \frac{\text{mean of experimental group} - \text{mean of control group}}{\text{standard deviation}}$$

**Table 2.2.2**  Cadmium concentration (mg kg$^{-1}$) in soils from two sites

|                    | Soil A | Soil B |
|--------------------|--------|--------|
|                    | 0.03   | 0.09   |
|                    | 0.05   | 1.5    |
|                    | 0.02   | 1.4    |
|                    | 0.01   | 0.98   |
|                    | 0.05   | 2.4    |
| Mean               | 0.032  | 1.27   |
| Standard deviation | 0.018  | 0.84   |

In our example of cadmium concentration we can insert our figures from preliminary experiments. Although soil A is not strictly a control group we can still use the standard deviation for this purpose as we want to determine the difference between the 'treated' group, which is soil B (herbicides applied), and the soil A, which did not have herbicide applied. The standard deviation used can either be that of the control group, or it can be a pooled value. In our case, our 'control' group is not strictly a control as it is simply another site, and therefore it is better to use a pooled value. It is important to note that this is not calculated by pooling the control and experimental data, but instead is an average of the two standard deviations from the two sets of data

$$\text{effect size} = \frac{1.274 - 0.032}{0.43}$$

which gives an effect size of 2.89, and means that we need to look for a difference in cadmium concentration of 2.89 mg kg$^{-1}$ dry weight between the two soils.

We can then insert these figures into a statistics programme such as Minitab and together with the suggested sample size (here the researcher suggests 20 samples) will then generate the power of the test, which in this case is one. This means that if you repeat the experiment many times then 100% of the time the null hypothesis will be correctly rejected. At no time will sampling error cause the null hypothesis to be accepted even though it should be rejected. If we alter the number of samples to two, then we obtain a power of 0.89, which now means that 11% of the time we may incorrectly accept the null hypothesis. In general a power of 0.80 or above is considered to be acceptable for statistical tests.

In addition to using the effect size, number of samples, and standard deviation to work out the power of your test, it is also possible to calculate the number of samples you require if you have the other three values (effect size, standard deviation, and the power). You can make a decision on the power of the test you require using 0.80 as a minimum. This is achieved using exactly the same component of Minitab, but inserting the power value instead of the number of values.

Table 2.2.3 provides a list of different values for standard deviation, effect size, and power. Input these into a statistical package such as Minitab and note how this affects the sample size required.

**Table 2.2.3** Influence of standard deviation, effect size, and power on sample size required

| Standard deviation | Effect size | Power |
|---|---|---|
| 25 | 5 | 0.80 |
| 10 | 5 | 0.80 |
| 10 | 1 | 0.80 |
| 10 | 10 | 0.80 |
| 10 | 10 | 0.90 |
| 10 | 5 | 0.95 |
| 25 | 5 | 0.95 |

By spending time ensuring that the sampling or experimental design is as robust as possible, it is much more likely that the data you collect will give meaningful results that stand up to scrutiny. However, it is also important to consider which statistical tests you may want to apply in order to test your hypotheses. This will ensure that sufficient data of the appropriate kind will be collected. Chapter 4 provides details of the key statistical tests you may need to use.

## 2.3    How reliable are your data?

Assume that you are studying the environmental impact of a factory that was thought in the past to emit a toxic pollutant. The factory has undergone recently some improvements designed to minimise emissions of that pollutant. You have data on the concentrations of that pollutant found in samples of air in a farm close to the factory for two dates: (a) 2 months before the improvements were implemented; and (b) 1 year after the improvements. At first glance the fact that the concentration in sample (a) of 10.2 ng m$^{-3}$ exceeds the concentration in sample (b) of 6.3 ng m$^{-3}$, seems to provide solid evidence that the improvements have been effective in reducing airborne concentrations of the pollutant around the factory. However, it is not as simple as that, and one has to take into account the precision, the reproducibility, or the uncertainty of the measurement. In the context of the current example, this means that a single measurement of concentration may vary by 30% (i.e. if one repeated the measurement it could vary by up to 30%), meaning that the concentration in sample (a) is 10.2 $\pm$ 3.06 ng m$^{-3}$ or between 7.14 and 13.26 ng m$^{-3}$; and that the concentration in sample (b) is 6.3 $\pm$ 1.89 ng m$^{-3}$ or between 4.41 and 8.19 ng m$^{-3}$. What this means is that there is overlap between the range of possible values for both samples, therefore we cannot say with certainty that the concentrations in the two samples are different and hence we cannot be certain that the improvements to the factory have been successful. Figure 2.3.1 illustrates this as Case 1.

If the uncertainty in measurement of the pollutant was lower – e.g. 10% – then as shown in Figure 2.3.1 as Case 2, the overlap in the range of possible concentrations no longer exists (the range of possible concentrations in sample A is 9.18–11.22 ng m$^{-3}$, while the possible concentrations in sample B lie between 5.67 and 6.93 ng m$^{-3}$) and we can say with certainty that the concentrations in the two samples are different and that the improvements to the factory have worked. Similarly, even if the uncertainty of measurement remains as high as 30%, one can still detect real differences between concentrations in samples A and B, if the difference is sufficiently large. As shown in Figure 2.3.1 as Case 3, if the concentration in sample A is 10.2 ng m$^{-3}$ compared with that of 4.6 ng m$^{-3}$ in sample B, then even with a measurement uncertainty of 30%, there is no overlap between the range of concentrations in sample A (7.14–13.26 ng m$^{-3}$) and those in sample B (3.22–5.98 ng m$^{-3}$).

To measure uncertainty (also known as reproducibility or precision), one has to conduct a series of repeat measurements (a minimum of five) on an identical sample.

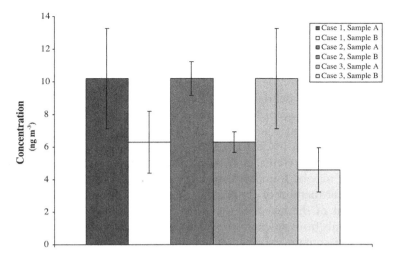

**Figure 2.3.1** Influence of reproducibility on comparison of pollutant concentrations before and after action designed to reduce concentrations.

For measurements in air, one needs to take a series of simultaneous measurements at – as far as is possible – the same location. Conversely, for media such as soil, sediments, plants, and water, one must make repeat measurements on identical sub-samples of a larger thoroughly homogenised sample. The uncertainty of measurement may then be expressed as the relative standard deviation (RSD) of the repeat measurements. The RSD is calculated as the standard deviation of the repeat measurements divided by the average of the repeat measurements (see section 3.1). It is then converted into a percentage by multiplying by 100. Note that for the usually small number of repeat measurements made (e.g. 5–10) the standard deviation is usually taken as $\sigma_{n-1}$ – Excel provides a value for this parameter as STDEV.

To illustrate, repeated analysis of the same soil sample gives the following concentrations of mercury (in $\mu g\ kg^{-1}$):

$$10.0; 9.8; 12.2; 9.4; 11.1; 13.6; 10.4$$

The $\sigma_{n-1}$ and average of these data are 1.50 and 10.93 $\mu g\ kg^{-1}$ respectively, and hence the RSD and thus the uncertainty of this measurement technique is 13.72%.

As well as producing reproducible data, it is essential that your data are accurate – i.e. correct. As a first step, measures must be taken to ensure that the equipment used to generate data on (e.g. concentrations of a pollutant in the environment) is accurate. At a very basic level, if one is weighing the amount of suspended sediment present on a filter through which a water sample has been filtered, one must ensure that when the balance tells you that the mass of suspended sediment + the filter = 213 mg, that this is correct, and not say 153 mg. To ensure accuracy of the balance, one calibrates it, by regularly checking that it gives correct readings for known weights (these are small discs of known weight – e.g. 10 mg, 100 mg, etc. – that can be purchased by a central reference laboratory such as the National Physical Laboratory in the UK).

In similar vein, more complex analytical instruments, such as mass spectrometers, must also be calibrated. Essentially, one must measure how the reading given by the mass spectrometer varies according to known amounts of a given pollutant (which can be purchased), so that when the mass spectrometer gives you a reading for a sample in which you wish to determine the amount of a pollutant, you know what amount that reading represents. A detailed discussion of how an analytical instrument is calibrated may be found as case study 1 in Box 5.1.1.

There is a further method of checking the accuracy of a measurement technique. This takes account of the fact that there are often extensive processes required to extract the pollutants from an environmental sample such as sediment, and purify the extract (which will contain many other chemicals in addition to the pollutants you wish to measure) before it is analysed on the analytical instrument. Although the analytical instrument itself is calibrated, the extraction and purification processes may result in losses of the pollutants, or even contamination, with the result that there is ample scope for recording inaccurate data even with an accurately calibrated instrument. To evaluate the accuracy of the entire measurement process, one analyses an aliquot of a standard reference material (SRM) or a certified reference material (CRM). These are samples of soil, sediment, etc., that have been thoroughly homogenised prior to being analysed by a number of the most reputable laboratories. These laboratories pool and compare their data on concentrations of various pollutants in the sample, and come to an agreement as to what the correct values of these are. The SRM/CRM is then made available with details of the correct or 'consensus' values together with an associated uncertainty (e.g. the consensus value for lead may be $8.34 \pm 2.1$ mg kg$^{-1}$), against which the values that you obtain can be compared, with the aim of recording concentrations that are as close as possible to the consensus values.

Of course, if you are relying on data that you yourself have not generated, then you cannot verify personally the accuracy and reproducibility of the data. However, wherever possible, you should ask the provider of the data what measures were taken to evaluate these vitally important parameters.

# 3

# How to summarise your data

## 3.1 Descriptive statistics

Once we have collected data we can use basic statistics to help us identify important patterns within the data and how much variation we have. By doing this we can make decisions as to which statistical tests are appropriate to apply to our data. Probably the most common and often the most useful basic statistic is the average value in the data, also referred to as the mean. This is simply calculated using the following equation:

$$\bar{x} = \frac{\sum x}{n}$$

This tells us to sum all the data values ($\Sigma$ is the symbol for 'sum') and divide by the number of values.

In Table 3.1.1 we have a set of measurements of plant growth with and without the addition of cadmium to the growth medium (experiment 1). Inspection of these data and in particular comparison of the mean values, suggests there is a difference between the two treatments and thus we may want to do further analyses. However, there is a drawback to the mean value. Let us take an alternative set of data from the same type of experiment repeated under identical conditions on another occasion (Table 3.1.2, experiment 2).

We can see that the mean values in experiment 2 are the same as those in experiment 1; but examine the data more closely. In experiment 1 the values that are obtained fall within a reasonably narrow range (30–39 and 15–21) and therefore we might suggest that the sample average is probably close to the actual population average. If we examine the second set of data, however (experiment 2, Table 3.1.2), we can see that the values range widely to the point of overlap (15–64 and 8–31) and there are many values that fall a long way from the average. In this case we may be more sceptical about whether the mean value really reflects the population mean as there is greater variation within the data. There is an alternative expression of the

*Student Projects in Environmental Science*     Edited by Stuart Harrad and Lesley Batty
© 2008 John Wiley & Sons Ltd

**Table 3.1.1** Growth of plants in the presence and absence of cadmium, experiment 1

| Shoot length (cm) | |
| --- | --- |
| Without cadmium | With cadmium |
| 35 | 16 |
| 30 | 18 |
| 36 | 20 |
| 34 | 19 |
| 39 | 18 |
| 35 | 18 |
| 35 | 21 |
| 34 | 15 |
| 30 | 16 |
| 31 | 16 |
| 32 | 17 |
| Mean[a] 34 | Mean[a] 18 |

[a]To two significant figures.

mean value that can be calculated, which can be used to measure the central tendency within sets of data that do not demonstrate normal distribution (see section 3.2). This is referred to as the geometric mean and is calculated using the following equation:

$$x = (x_1 x_2 x_3 x_4)^{1/n}$$

**Table 3.1.2** Growth of plants in the presence and absence of cadmium, experiment 2

| Shoot length (cm) | |
| --- | --- |
| Without cadmium | With cadmium |
| 40 | 8 |
| 15 | 17 |
| 16 | 22 |
| 64 | 31 |
| 38 | 11 |
| 37 | 14 |
| 58 | 26 |
| 23 | 21 |
| 18 | 18 |
| 34 | 14 |
| 27 | 14 |
| Mean[a] 34 | Mean[a] 18 |

[a]To two significant figures.

What this tells us to do is to multiply the data values and then take the $n$th root of this figure where $n$ = number of data values. So for example if we take the figures for shoot length in the absence of cadmium from Table 3.1.2:

$$x = (40, 15, 16, 64, 38, 37, 58, 23, 18, 34, 27)^{1/11}$$

Then:

$$x = 30.2$$

As with the arithmetic mean, this and the other statistical terms described in this section are calculated easily by software such as Excel, Minitab, and SPSS.

Apart from the mean value there are two other basic statistics that can be calculated. First there is the median value. This is simply the central (or the middle) value if all the data values are arranged in ascending order of magnitude. So in the examples above the two median values in experiment 1 are 34 and 18. In this case both of the median values are the same as the average values. What about the second set of data (i.e. from experiment 2)? In this case the median values are 34 and 17. Again the medians are close to the average values, we will return to this point in section 3.2. Note that if there are an even number of data values there will be no central value. In such cases, to calculate the median, we take the two middle values, sum them and divide by two. Finally we can calculate the modal value, which is simply the value that occurs most often (the **mode**). So in the first set of data the modal values are 35 for the treatment with cadmium. In the other treatment however you will notice that there are two values that occur three times, 18 and 16. In this case the data is referred to as **bimodal**. In the second set of data the modal value for the treatment with cadmium is 14, whereas in that without cadmium there is no value that occurs more than once, and therefore there is no mode. The problem with the mode is that it is easily changed by the influence of just one value. In large datasets the likelihood of more than one modal value increases and thus its usefulness rapidly dwindles.

Once you have calculated the median you can then describe the data distribution in another way. Any point along the ranked distribution of values can be referred to as a **percentile**. This is a descriptive tool that divides the dataset into divisions also known as **quantiles** relative to 100. Therefore the 75th percentile (upper quartile) will relate to the data value that is 75% or three-quarters of the way up the ranked list of data values and the 50th percentile is the same value as the median. We can thus refer to a single data point in relation to all others, for example if a plant is equal to or greater in height than 67% of other plants in the population then we can say that it is in the 67th percentile of heights in the population. Calculate the 90th percentile value of plant shoot length in the presence of cadmium from the data in Table 3.1.1. You should find it to be 20.

All these calculations give some indication about the dataset that has been calculated but on the whole provide very limited information. Earlier we mentioned

the importance of knowing the variation within a dataset and there are basic statistics
that we can use to provide details of this variation.

The first thing that we can determine is the extent of the variation around the
sample mean, which is referred to as the **standard deviation** (SD). As mentioned in
section 2.3, where the number of data points are relatively small (typically less
than 10), then it may be expressed as

$$\sigma_{n-1} = \sqrt{\frac{\sum (x - \bar{x}^2)}{n - 1}}$$

The mean value of the data is first calculated and then this is subtracted from each of
the numbers in the dataset and the result squared. These results are then summed and
divided by the number of data points $-1$ (e.g. if there are 10 data points, then you
divide by 9 – see also section 2.3). The square root of this value is then calculated.
This value is referred to as $\sigma_{n-1}$.

Once the number of data points exceeds 10, then it is more appropriate to use $\sigma_n$.
This is identical to $\sigma_{n-1}$, except that the denominator is the number of data points
rather than the number of data points less one.

$$\sigma_n = \sqrt{\frac{\sum (x - \bar{x}^2)}{n}}$$

As you can see this can be a very laborious process in large datasets, but fortunately
both $\sigma_n$ and $\sigma_{n-1}$ can be calculated easily on any statistics package. For example in
Excel, $\sigma_n$ and $\sigma_{n-1}$ may be generated using the STDEVP and STDEV functions
respectively. The standard deviation (SD) value is a measure of the confidence you
have that a particular data value will fall within a particular range (the mean $\pm$ SD).
In normally distributed data (see section 3.2 for an illustrative figure) 95% of the data
values will fall within the region defined as mean $\pm$ 2SD.

Try to calculate the SD for the datasets in Tables 3.1.1 and 3.1.2. You should obtain
the following results to two decimal places (Table 3.1.3). Now we can see clearly
that the variation in the second dataset (experiment 2, Table 3.1.2) is much higher
than that in the first (experiment 1, Table 3.1.1), and this supports our suggestion
earlier that the mean value for this set of data is not very representative.

**Table 3.1.3**  Standard deviation ($\sigma_n$) for datasets in Tables 3.1.1
(experiment 1) and 3.1.2 (experiment 2)

| Experiment | Without cadmium | With cadmium |
|---|---|---|
| 1 | 2.63 | 1.77 |
| 2 | 15.5 | 6.44 |

An additional measure that is sometimes calculated for datasets is the **relative standard deviation**, sometimes referred to as the **coefficient of variation** (see also section 2.3). If we knew that the standard deviation for the height of beech trees was 100 cm and that for rhododendron bushes was 10 cm then we would come to the conclusion (correctly!) that the height of beech trees is much more variable than rhododendron bushes. However, this does not tell us how this variation relates to the actual size of the organism. The relative standard deviation allows us to determine this by dividing the standard deviation by the mean. So for example if the mean height of beech trees is 2100 cm and that for rhododendrons is 86 cm then the calculation is as follows

$$\text{beech trees} : 100/2100 = 0.05$$
$$\text{rhododendron bushes} : 10/86 = 0.12$$

These values can also be presented as percentages – i.e. 5% for beech trees and 12% for rhododendrons, telling us that relative to their size rhododendrons are more variable.

The SD figure (and the relative SD) is useful but the mean value we use here is one based upon the data that you have collected, so for example if you collected only five samples then the mean value around which variation is considered is based upon only these five samples. If you repeated the data collection then the mean value (and the variation around it) would probably be very different. What we want really to know is whether the mean of our population is **representative** of the true population mean. If we repeated the data collection many times we would obtain a series of mean values which when plotted would be normally distributed (see section 3.2 for further discussion of normal distribution). We can then find the mean of the means and the standard deviation of it, which is known as the **standard error** (SE). Fortunately there is a statistical way to estimate SE using the equation below.

$$SE = SD/\sqrt{n}$$

The larger the standard error the more chance there is that your measured mean is greater or smaller than the true population mean. Let us return to our data from Tables 3.1.1 and 3.1.2.

The first set of data had two mean values of 34 and 18 and we suggested on the basis of visual inspection that the variation in the data was fairly small. If we calculate the standard error in the data then we obtain values of 0.83 and 0.56. The way that this would be reported would be $34 \pm 0.83$ and $18 \pm 0.56$. Now if we examine the second set of data that we suggested, despite having the same mean values, had a greater variance, then we obtain values of $34 \pm 4.91$ and $18 \pm 2.04$. This gives us a good indicator that in our second set of data our measured mean is more likely to be different to the true population mean, and we may not be so confident about conclusions we draw from our data.

## 3.2  Probabilities and data distributions

Statistics is essentially all related to probability. What we are trying to do is to determine whether patterns within the data we have collected are there by pure chance or are 'real' patterns. For example, say we have made a series of measurements of pH and iron concentration in river water and obtained the following relationship between the two variables (Figure 3.2.1).

From Figure 3.2.1 it appears that there is what is termed a negative relationship between the two variables, in that as pH increases iron concentration decreases. However, how do we know that this apparent trend is not a result of chance? For example if we went out and repeated the survey, how do we know that we would not get a completely different set of data that would not show this trend (we saw in section 3.1 how variable different datasets for the same experiment on different occasions can be). This is where statistics comes in. In this instance, statistics enables us to determine the likelihood (or probability) that the trend that is evident in our data occurred by chance – and by inference, that it did not and is therefore a genuine trend. In most of the natural sciences it is common practice to take 5% as the level at which we decide whether a relationship has occurred by chance or not. If the probability that it has occurred by chance is less than 5% then we can accept the relationship as being significant (i.e. we are 95% confident that it did not occur by chance). In other disciplines the significance level may be different, for example in medical sciences it is sometimes stricter and given as 1%. When statistical tests are applied, the results given will include a probability value presented as a **p-value**, which is usually given as a proportion rather than a percentage. So in our example

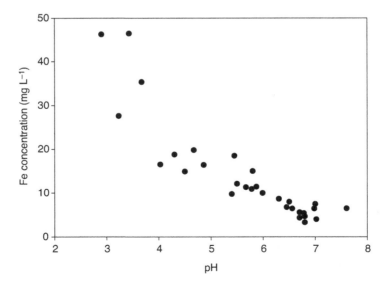

**Figure 3.2.1**  Scattergram showing the relationship between pH and iron concentration (mg L$^{-1}$) in river water.

above the *p*-value for the correlation (see section 4.2 for more on testing for relationships) is given as $p < 0.05$ telling us that the probability of obtaining the trend in the data is less than 5% and therefore is a real relationship and not one that occurred by chance.

The use of statistical tests enables us to establish whether relationships and differences that are evident in our data occurred by chance or not. However, statistical tests are derived from mathematics and they make several assumptions about the datasets that are being tested. If our datasets do not fit these assumptions then we cannot validly use the statistical test (we will return to this in section 3.3). One of the main assumptions of many of the tests relates to the 'distribution' of the data. Many natural populations, when measured give a typical distribution plot shown in Figure 3.2.2. This typical 'bell' shaped curve is known as the Gaussian Curve and the shape known as a **normal distribution**.

From Figure 3.2.2 we would expect the mean, median and mode to lie very close together and if we calculate these values this is proven to be true (mean = 5.64, mode = 6 and median = 6). However the fact that the mean, median and mode are close does not always indicate normal distribution. For example, data values may be very scattered or concentrated and in these cases the distribution is not normal even though the mean, median and mode are close together. Look back to the example used in section 3.1 (experiment 2, Table 3.1.2), where the mean and median were both very close but by simply looking at the data we can see that it is not normally distributed. Plotting a histogram of these data confirms this (Figure 3.2.3).

If the majority of scores fall mainly below the mean then the distribution is referred to as being **positively skewed** (Figure 3.2.4A) and if they fall mainly above the mean then it is **negatively skewed** (Figure 3.2.4B). The mean value in

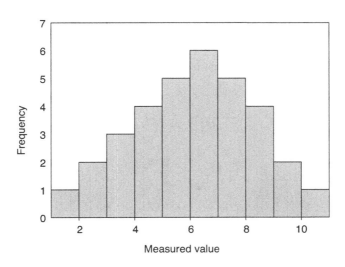

**Figure 3.2.2** Histogram of data showing normal distribution.

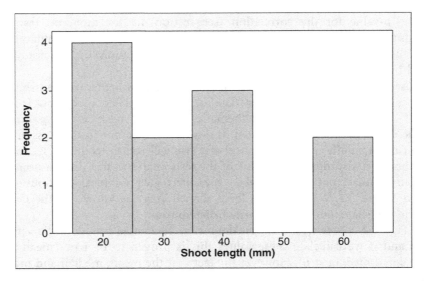

**Figure 3.2.3** Frequency distribution of data collected on shoot length in response to cadmium supply (see Table 3.1.2, values without cadmium).

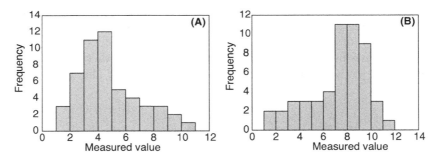

**Figure 3.2.4** Frequency distributions showing positive skewness (A) and negative skewness (B).

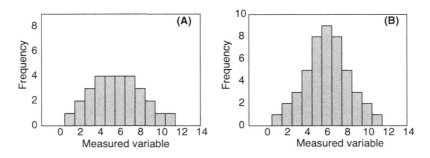

**Figure 3.2.5** Frequency distributions showing platykurtosis (A) and leptokurtosis (B).

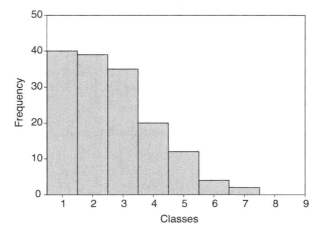

**Figure 3.2.6** Log-normal distribution of the relative abundance of moth abundances captured within aerial traps.

Figure 3.2.4A is 4.31 and in Figure 3.2.4B is 6.78, but it is clear from the graphs that this does not accurately reflect the data as a whole.

Data distributions can also show something known as kurtosis where the normal bell curve is slightly flattened (platykurtosis) or peaked (leptokurtosis) (Figure 3.2.5).

We will come back to the importance of data distribution in section 3.3.

In some cases natural populations would not be expected to show normal distribution but instead show a log-normal distribution, which is a skewed distribution. For example species–abundance relationships have been shown to be log-normally distributed where the total number of species is large, therefore it applies to many biological communities (although recent research has questioned this). Where the variable $x = \log(y)$ the data are log-normally distributed. Figure 3.2.6 shows a typical dataset with this type of distribution.

Although frequency distributions are useful in identifying very obvious patterns in data, it may be difficult to judge exactly when data are normally distributed using this method. There are statistical tests that can be used to test whether data are normally distributed, which we will return to in section 3.3.

## 3.3  Choosing the appropriate statistical test

Before we can select the appropriate statistical test to apply to our data we need to examine the distribution of the data. This is because when statistical tests are derived, statisticians make several assumptions about the data to be tested. If these assumptions are not met, then the tests will not be mathematically sound and the outputs misleading.

Probably the most important of these assumptions is whether the data are 'normally distributed' or not. Only normally distributed data can be evaluated by

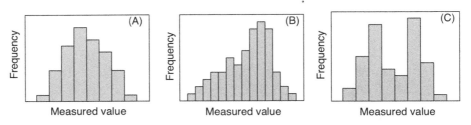

**Figure 3.3.1** Frequency distributions showing normal distribution (A), negative skewness (B) and bimodal distribution (C).

parametric tests. If you cast your mind back to section 3.2 you will remember that based upon probability and natural populations, a 'bell' shaped curve was derived known as a Gaussian Curve (Figure 3.2.2). This describes data that are normally distributed, but you will also remember that often data do not fit this pattern and may be skewed, bimodal, or log-normal. So how does one determine whether the data are normally distributed? Section 3.2 described the visual inspection approach, which is simply to plot a histogram of the data values and 'eyeball' it. Do the data look normally distributed? Figure 3.3.1 gives some examples of histograms generated from datasets. It is evident that the data in A looks to be normally distributed whereas the data in B and C do not.

Of course 'eyeballing' is subjective and if your statistics are going to stand up to scrutiny then it is essential to test the distribution mathematically. This can be done easily by using statistical tests that compare a data distribution against a theoretical distribution. There are a few tests available to us, the main ones being the chi-squared and the one-sample Kolmogorov–Smirnov. It should be noted, however, that these tests can be carried out only on datasets that are sufficiently large, typically $> 50$ values. Remember that if you are comparing data from two populations then each dataset must be tested for normal distribution.

You may remember from section 3.2 that many biological communities do not fit the normal distribution but instead show log-normal distribution. There are tests available that can be used to determine whether the data differ from this type of distribution, just as for normal distribution.

Once the nature of the distribution has been determined we can make a decision as to whether parametric tests may be used. What happens if the data are not normally distributed? There are two alternative approaches that we can take: the first of these is to simply abandon parametric testing. There is an entire suite of tests that are designed specifically to test for differences and relationships when data are non-parametric.

**Table 3.3.1**  Caterpillar length (mm) on two different shrubs

| Shrub A | 55   | 56 | 60 | 68 | 120  | 53 | 60.5 | 56   | 60 | 112.5 | Mean = 70.1  |
|---------|------|----|----|----|------|----|------|------|----|-------|--------------|
| Shrub B | 54.5 | 55 | 64 | 60 | 65.5 | 58 | 54   | 55.5 | 61 | 63    | Mean = 59.05 |

**Table 3.3.2** Caterpillar length (log-transformed) on two different shrubs

| Shrub A | 1.74 | 1.75 | 1.78 | 1.83 | 2.08 | 1.72 | 1.78 | 1.75 | 1.78 | 2.05 | Mean $= 1.83$ |
|---------|------|------|------|------|------|------|------|------|------|------|---------------|
| Shrub B | 1.74 | 1.74 | 1.81 | 1.78 | 1.82 | 1.76 | 1.73 | 1.74 | 1.79 | 1.80 | Mean $= 1.77$ |

There is nothing wrong with this approach although the tests are less powerful than the parametric equivalents.

The other alternative is to transform the data and a number of options are available here. Probably the most commonly used are the logarithmic function ($\log_x$) or the square root function. These simply calculate the log or square root of each data value. By doing this we are reducing the disproportionate influence of unusual values within the dataset (either very large or very small values). To illustrate this, Table 3.3.1 lists the data obtained when we measured the length of caterpillars on two different shrubs.

The data from the two shrubs does not seem to vary greatly yet the mean value for shrub A is a lot higher than shrub B. This is because the dataset is greatly affected by the presence of two very large caterpillars measured as being over 110 mm long. It may be tempting to remove these values but this would invalidate the whole dataset as there is no reason to suspect they are false readings and therefore if we were to remove them, we would be effectively making the data up. Instead we can transform the data using $\log_{10}$ which gives us the transformed dataset in Table 3.3.2.

By transforming the data we have reduced the influence of the extreme values and the two means are now much more representative of the individual values within the dataset. The transformed data can then be replotted and tested for normality again. Of course, if these transformations do not solve the problem then the only option is fall back on non-parametric testing.

Once we have decided whether our data are amenable (with or without transformation) to parametric testing, we can apply the appropriate statistics to test our original hypothesis – see Chapter 4. Remember that this should have been thought out before the data collection aspect of the research project and will depend upon whether you want to establish differences or relationships and the type of data you will be collecting (see section 2.2).

# 4

# Testing hypotheses

## 4.1 Coincidence or causality?

So far we have considered a number of aspects of research including experimental design and some preliminary analysis of the data. However, before we move on to more advanced statistics designed to establish relationships and differences (see section 4.2) we need to cover one important concept. Whenever we set out to undertake research we are in essence trying to understand the fundamental processes within the environment by identifying relationships and patterns within data we have collected. However, even if we establish whether there is a relationship between different variables, we do not know whether there is a **causal** link between the two. An obvious example (if flippant) is that ice cream sales correlate positively with concentrations of PCBs in air (i.e. as one increases so does the other – see section 5.1), but one does not cause the other (it is high temperatures causing both)! A more complex example is that we may have hypothesised that algal growth decreases with a decrease in the environmental pH as a result of some preliminary observations. We have then undertaken an experiment to determine whether this relationship occurs, and found that it does. However, although we have established that there is a relationship between the two variables we do not know necessarily that the change in pH was the direct cause of the reduction in algal growth. We know from basic chemistry that many toxic metals become increasingly bioavailable when pH decreases, and therefore it is also conceivable that the increased availability of the metals to the algae caused a toxic effect and a reduction in growth. Whether this is true or not cannot be determined from the relationship data.

This is often a problem in interpreting data, particularly when dealing with real environmental data from the field where there has been limited control on the variables. So how do we know when a relationship is causal or not? There is no absolute answer to this question, however there are a number of indicators that we can take into account when drawing conclusions from the data. In 1965 a set of criteria were proposed by Bradford Hill by which causality between data may be assessed. He

*Student Projects in Environmental Science*   Edited by Stuart Harrad and Lesley Batty
© 2008 John Wiley & Sons Ltd

considered these criteria in relation to epidemiology and the causes of disease but these criteria are also useful when considering other environmental data. There has been much criticism and discussion of these criteria within the literature but the important thing to remember is that the criteria are not a 'check list' by which relationships are decided but is a set of indicators that can be used to support the presence of a causal relationship. Essentially we can distil Bradford Hill's work into nine criteria.

### 4.1.1   The nine criteria of Bradford Hill

These criteria can be used in the decision making process as to whether a casual relationship can viably be proposed as an explanation for a relationship. However, we need to emphasise that this is not a checklist. Any one of these criteria may be used to support a causal hypothesis, and even if all criteria are met, this does not always lead to complete acceptance of the hypothesis. Arguably it is experimental evidence that will provide the strongest support, but not in every case and it is not always possible to carry out suitable experiments. However, well designed sampling strategies and experiments which involve consideration of statistics will maximise the possibility of supporting (or otherwise) a causal relationship.

*Strength*
Bradford Hill used the death rate from lung cancer amongst cigarette smokers to illustrate this point, stating that the death rate among smokers is 9 to 10 times that of non-smokers. What he is trying to communicate here is that the strong relationship between the two is indicative of a causal relationship. We can use the example of extinction rates of species to illustrate this point. If we examine the increase in extinction rates with the increase in human population then we find that there is a significant difference in comparison to background rates (Figure 4.1.1.). This relationship may

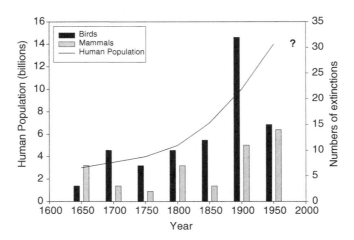

**Figure 4.1.1**   Relationship between human population and mammalian and bird extinction rates.

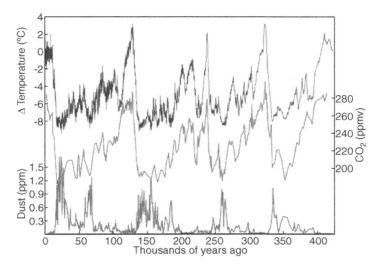

**Figure 4.1.2** Data from the Vostok Ice Core showing the relationship between $CO_2$ in the atmosphere and global temperature [Reproduced from wikipedia.org].

suggest that the increase in human population may **cause** extinction of species, but of course it does not tell us anything about the processes involved, which may be numerous.

*Consistency*

This essentially refers to whether the relationship is found by a number of different people, at different times and in different places, which if so increases the likelihood of a causal effect. So as an example let us take the classic case of $CO_2$ and temperature (Figure 4.1.2).

It has been suggested that there is a causal relationship between $CO_2$ and global temperature that is supported by evidence from ice cores that demonstrate that the relationship has occurred at numerous times over the past 400 000 years. This is a good example of a possible causal relationship as not only do we have consistent data but the relationship is strong and can also be explained scientifically. However, if you examine the literature surrounding this relationship you will still find that a causal link is disputed by some researchers, demonstrating the problems with trying to determine whether the relationship occurs by chance or whether there is causality.

*Specificity*

In cases where the cause and effect occurs in a specific population in a specific location and there is no other obvious explanation then this is thought to increase the probability of a causal relationship. An example of this may be found at Sudbury, Ontario. Within the vicinity of a copper smelter significant ecological damage was found in the guise of vegetation death and contamination of local water bodies. The maximum ecological damage was found close to the smelter and decreased with distance from the main site. A direct causal relationship was proposed and established using other supporting evidence.

*Temporality*
This criterion is really the equivalent to the chicken and egg scenario. So if we return to our carbon dioxide example, does the increase in carbon dioxide cause a rise in temperature or does the rise in temperature (due to some other factor) cause a rise in carbon dioxide. In science we need to be careful when interpreting datasets that may have this temporal problem, but fortunately they are fairly rare.

*Biological gradient*
When examining relationships between variables that include some kind of biota we may wish to look for evidence of a biological gradient or dose–response curve. For instance, in response to an increase in pollutant concentration, the numbers of a particular organism may decrease linearly. These kinds of relationships often do not have simple linear relationships and may follow some other pattern. For example the growth rate of bacteria with time shows a non-linear relationship (Figure 4.1.3), and this is often seen in other biological populations in response to time or other factors such as temperature. It is important to look for these types of relationship within biological data and more information can be found in section 5.3.

*Plausibility*
Arguably the key thing that one should consider when examining relationships is whether it makes sense, i.e. can the relationship be explained scientifically and does it

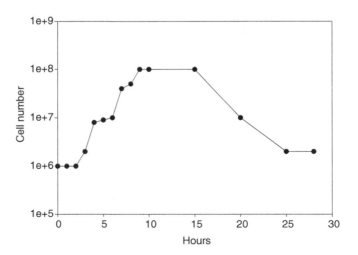

**Figure 4.1.3**  Bacterial growth curve.

conflict with our existing knowledge of the subject? We should already have some idea about this from the existing literature (which we would have used to generate our hypothesis in the first place) and therefore if we cannot provide a plausible explanation for the relationship we may want to question it. However, a word of caution here, just because there is no obvious explanation does not necessarily mean that there is not a causal relationship. Remember that we carry out research to increase our understanding because we do not have an explanation for everything! For example, we may have found that as the population of the common rat increases so does the occurrence of the fruit fly. We know from previous knowledge that there is no reason why an increase in rats would directly cause an increase in fruit flies and so we would be more inclined to look for a common factor that we have not measured, such as the amount of rotting food.

Bradford Hill actually suggests that plausibility is not a viable criterion by which to judge the possibility of a causal relationship. He argues that what is plausible depends upon the scientific knowledge of the day and that just because a relationship may not be explained, does not mean that it does not exist. There are many examples of patterns and relationships in science that have been observed, explanations proposed and dismissed as being too radical or contradictory to the current accepted science. However, further collection of scientific evidence has sometimes led to major changes in our understanding of the natural world (continental drift and the theory of evolution are two examples that immediately spring to mind). Therefore we need to make sure that we do not dismiss relationships, simply because an immediate explanation does not present itself, but instead further investigation may be warranted.

### Coherence

This criterion relates to that of plausibility. We have observed that the limitations of our scientific knowledge may prevent our acceptance of a relationship, simply because we cannot put forward a plausible explanation. Bradford Hill suggests that a proposed explanation or indeed an observed relationship should 'not seriously conflict with the generally known facts.' We may argue at this point that many previous advances in science have actually conflicted with generally known facts but there is a balance that needs to be struck between patterns and explanations that 'fit' or are coherent with previous data and those that are entirely at odds with known facts.

### Experiment

Once a relationship has been observed and a cause and effect hypothesis proposed, experimental evidence may be used to support the hypothesis. For example, the observation that an increase in eutrophication (an increase in nutrient concentration) within rivers correlates with an increase in algal growth may suggest a causal relationship. This can then be tested further using controlled experiments in the laboratory where nutrient concentrations are varied and other variables (e.g.

temperature, pH) are kept constant (see section 2.2 for more information on experimental design) to determine whether this the is cause of the increase in algal growth. Experimental evidence is arguably the best way of providing support for a causal relationship.

*Analogy*

It is possible in some circumstances to use the existence of another causal relationship to support a similar relationship by analogy. For example the effect of one heavy metal on the growth of a plant species may be used to support other less convincing evidence of a similar response to another heavy metal. We need to be very careful here to ensure that the analogy is justified and it could be argued that this is unlikely to be accepted generally without other supportive evidence.

## 4.2   Relationships and differences

In essence there are two questions that we can answer within research: first, whether there is a relationship between two or more variables; and second, whether there is a difference between two or more populations. Sometimes we wish to answer only one of these questions, and sometimes we need to answer both. Whatever the situation, it is absolutely essential to know which of these you are trying to determine with your data collection before you design your experiment, otherwise it is easy to collect the wrong type or amount of data.

So let us say that we are interested in the uptake of metals from soils by a particular plant species. The first hypothesis that could be generated would be:

Concentrations of zinc in plants are influenced by concentrations of zinc in soils

If you remember from our discussion of research project design (section 2.2) we have a number of options open to us. We may decide to carry out a sampling campaign, taking samples of a particular plant species from a number of sites containing different concentrations of zinc. In this case we then analyse zinc concentrations in plants and soils and compare them in order to establish whether the two are related. Alternatively a simple experiment can be carried out by growing plants in a range of soil zinc concentrations and then analysing the plant tissues for zinc. We obtain the same type of data from both approaches but in the latter experiment we have much more control over the zinc concentrations and any confounding factors. Figure 4.2.1 illustrates the relationship that we may get from the two different approaches.

From Figure 4.2.1A it is evident that there is a positive relationship between the two variables, although it appears that it may not be linear. We can test this using a correlation coefficient (section 5.1). The laboratory approach provides a very different set of data, although again there appears to be a positive relationship between soil and plant zinc concentrations. The difference here is that we have controlled one of

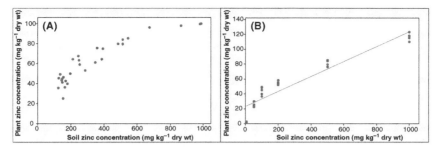

**Figure 4.2.1** Scatterplots of zinc concentration in soils and plant tissues (mg kg$^{-1}$ dry wt) from (A) a field sampling campaign and (B) a laboratory experiment.

the variables and therefore we can use the relationship to predict the response of the other variable (more details of this can be found in section 5.1). So from both types of approach we can see that there is a direct relationship between the two variables. One common mistake when dealing with this kind of data is to assume that if there is a correlation between the two variables then this means that the two measurements are the same. This may be (and probably is) not the case. Look carefully at Figure 4.2.1 in which we established there was a positive correlation. Note that soil zinc concentration is much higher than the equivalent plant zinc concentration. We can see in 4.2.1B that when soil zinc concentrations are 1000 mg kg$^{-1}$ the concentrations of zinc in the plants are only around 120 mg kg$^{-1}$, i.e. they are not the same. Let us take another example to illustrate this point.

In this case we have measured benzene concentrations in both indoor and outdoor air from a number of locations. Figure 4.2.2 shows that there is a relationship between the two and a Pearson's correlation test shows us that this is statistically significant at $p < 0.05$. However, when outdoor air is high (11 μg m$^{-3}$) we can see

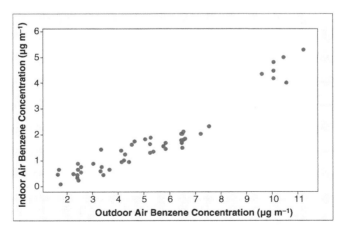

**Figure 4.2.2** Benzene concentrations (μg m$^{-3}$) measured in indoor and outdoor air.

that the equivalent indoor air is not at an equivalent concentration but is instead at levels of around $5\,\mu g\,m^{-3}$. We may hypothesise that outdoor air is the source of benzene in indoor air because the two measurements are correlated, but concentrations in outdoor air exceed those indoors.

So we can see that correlations may allow us to determine a relationship between two variables but what we now need to ask is has a change in one variable **caused** a change in the other? Refer back to section 4.1 if you are uncertain about the answer to this question.

In the examples so far we have been trying to establish whether there is a relationship, but what if we want to establish whether there is a difference between the two variables. In the example above we noted that outdoor air concentrations of benzene were higher than those in indoor air, so we may now want to determine whether there is a significant difference between the two. This would involve simply reanalysing the data using an appropriate test (see section 4.3.1).

In the case of the plant uptake examples, we may want to establish whether there is a significant difference between plant species in their uptake of zinc from soil? In this case our hypothesis would be:

Plant species A takes up greater amounts of zinc from soils than plant species B

This would now involve a different type of experiment in which we could grow a range of plant species in soils containing the same amount of zinc and then measure the uptake of zinc into the plant tissues. The results for such an experiment are shown in Figure 4.2.3.

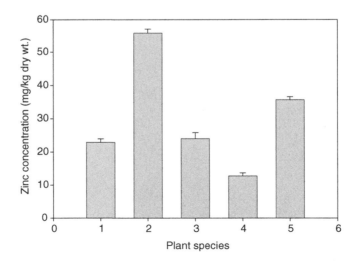

**Figure 4.2.3** Zinc uptake in the plant species *Juncus effusus* (1), *Brassica juncea* (2), *Typha latifolia* (3), *Festuca arundinacea* (4) and *Phragmites australis* (5) grown in soils containing $100\,mg\,kg^{-1}$ dry wt zinc. Bars indicate standard error ($n = 5$).

In this case we are not looking for a direct relationship between the soil and plant zinc concentrations but rather differences in the populations (the different plant species). We can see clearly from Figure 4.2.3 that there appears to be a difference in zinc uptake between the different species, and we would be able to test this further for its statistical significance by using appropriate statistical tests (section 4.3).

## 4.3 Testing for differences

When we are trying to establish whether two or more populations differ from each other in some way (e.g. that one river is more polluted with lead than another) then we have a range of statistical tests available to us. The choice of the appropriate test is made with consideration of whether the data are normally distributed (see section 3.2), how many groups we are interested in, whether there is one dependent variable or more than one, and the type of data we have. The statistical tests designed for data analysis use a variety of mathematical models and make assumptions about the data that are being analysed. One of the key assumptions in those that are termed **parametric** tests is that the data are normally distributed (see section 3.3), and there are then a number of different tests available depending on the other factors (e.g. number of groups, number of variables, etc.). If the data are not normally distributed (or **non-parametric**) then an alternative set of tests are available. Alternatively, as shown in section 3.3, the data may be log-transformed to facilitate parametric testing.

### 4.3.1 Parametric tests

First we shall start with the situation where we have made a set of measurements of two groups (or populations). If we want simply to determine whether there is a significant difference between the two groups then we can apply something known as a student t-test (often just referred to as a t-test). There are two different types of t-test available to us, first the independent t-test which is used for samples that are truly independent and second the paired t-test. The paired test is used where each measurement in one group can be linked and therefore directly compared with a particular measure in the other group. Let us take an example where we want to determine changes in phosphate concentrations within rivers to illustrate this concept. If we want to compare phosphate concentration at location 1 on a number of days with phosphate at location 2 (which is within the same river) on the same days then we can justifiably use the paired t-test. This is because the two samples would be expected to be related. However, if we wanted to compare two locations that were in **different** rivers on the same days then we would use the independent t-test because the samples are not related in any way. As a general rule paired samples will require more replication than an independent t-test.

There are a number of assumptions that need to be met before a t-test can be applied. First the data must be parametric, the samples must be random, and variance within the two groups must be similar. In addition the independent t-test must only be used for samples that are randomly assigned to each group. Thus it is not suitable for time-series data, individuals grouped in one site or repeated samples obtained from the same test subject. The advantage of using t-tests is that they can be applied to raw count data but also for other types of data including constant interval data (see section 2.1). In addition the independent t-test can be applied to unequal sample sizes in the two groups. However, be aware that the t-test can only be used to compare **two** groups; other tests are available to compare more than two groups. We will deal with these later in this section.

For example, if we have hypothesised that there is a difference in the concentration of a particular pollutant in two environments (indoor and outdoor air) and collected 50 samples from each environment, then we can use an independent t-test to determine whether this difference is significant (i.e. is indoor air significantly more or less contaminated than outdoor air with a given pollutant?), as long as the data are normally distributed. Table 4.3.1 provides a sample set of data for analysis.

Start by testing whether the two datasets are normally distributed using a one-sample Kolmogorov–Smirnov test. This is accomplished using either Minitab or SPSS. You should find that both sets of data are normally distributed. This means that we can proceed with the t-test.

A t-test carried out on the data in Table 4.3.1 using Minitab 14 gives the following output

$$\text{T-Value} = 4.63 \quad \text{P-Value} = 0.000 \quad \text{DF} = 75$$

The t-value is the main output from a t-test and reflects how large the difference between the sample means is. The greater the t-value the greater the difference between the sample means, and the smaller the variance around the sample means. The t-value would previously have been looked up in statistical tables to determine the equivalent p-value which can tell us whether there is a significant difference or not; however, statistical packages now give us the value, which in the example above is $< 0.001$ ( do not forget that statistical significance can also be $< 0.01$ or $< 0.05$). Therefore, there is a significant difference between the concentrations of the compound measured in indoor and outdoor air. The other value that is important

**Table 4.3.1**   Benzene concentrations ($\mu g\ m^{-3}$) measured in indoor and outdoor air samples

| | |
|---|---|
| Indoor air | 1.30, 2.40, 0.46, 6.50, 2.01, 2.15, 1.46, 10.53, 1.20, 2.56, 6.45, 4.53, 5.85, 5.75, 6.35, 10.04, 9.58, 5.23, 4.12, 2.35, 1.52, 5.36, 5.86, 6.45, 4.21, 1.25, 7.20, 10.05, 2.40, 3.65, 3.30, 4.42, 2.24, 1.01, 10.42, 11.21, 10.04, 5.25, 5.04, 4.25, 4.32, 3.33, 6.60, 2.54, 4.10, 0.09, 1.14, 6.53, 4.63 |
| Outdoor air | 2.45, 1.03, 5.24, 1.32, 3.42, 3.02, 3.36, 4.02, 2.42, 1.01, 0.88, 0.24, 1.42,2.12, 7.52, 1.02, 1.99, 1.65, 2.94, 2.43, 1.68, 1.02, 3.33, 3.52, 2.45, 6.53, 1.01, 1.42, 1.82, 3.44, 1.31, 1.02, 0.45, 1.65, 6.42, 0.78, 1.46, 2.22, 3.13, 2.50, 6.46, 2.12, 0.58, 1.65, 1.97, 1.74, 3.41, 2.46, 2.12 |

here is the DF value which is an abbreviation for **degrees of freedom.** This value is related to the size of the sample and the number of groups that the data are assigned to for comparison. For every measurement you make there will be a degree of freedom, and then each test you use to answer a specific question will use up these degrees of freedom. For simple tests only one degree of freedom is used and therefore the DF value will be $n - 1$. However, when you start carrying out more advanced tests such as ANOVA (ANalysis Of VAriance), more degrees of freedom will be used up. The problem is that if the DF value is 0 then the tests will not work. This can occur if you have insufficient sample sizes for answering a number of questions.

There may be instances where we do not simply want to determine the differences between two groups, but instead we want to identify whether the results from one population match an expected set of results (this could also be referred to as 'goodness of fit'). In this case we need to have an expected result and this can either be determined from probability or derived from pre-existing datasets.

In a situation where we have two groups and counts have been made of the number of observations that fall within each group, then we apply a test known as chi-squared (denoted as $\chi^2$). This test cannot be applied to constant interval data (see section 2.1) or derived values. This test does not make assumptions about the normality of datasets, indeed it can be used to test for normality, and therefore is not strictly a parametric test but can be applied to both parametric and non-parametric data.

The overall equation for the chi-squared test is given below and you can see that it is simply a way of comparing observed against expected values. This can be calculated automatically by most statistical packages.

$$\chi^2 = \sum (O - E)^2 / E$$

where $O$ is the observed number and $E$ the expected number.

For an example, let us suppose that we want to determine the preference of a particular insect parasite for four different plant species. We could carry out a simple laboratory experiment that allows the parasite to choose plant species, and simply count the occurrence of the parasite on each species. What we do is to base our expected results on probability, in that there is a 1 in 4 or 25% chance that the parasite will be found equally on each of the plant species. This can be illustrated effectively in a table where we can compare our actual results with the expected results (Table 4.3.2).

**Table 4.3.2** Observed parasite counts on different plant species and theoretically expected results

|  | Plant species | | | |
|---|---|---|---|---|
|  | 1 | 2 | 3 | 4 |
| Observed | 12 | 44 | 26 | 18 |
| Expected | 25 | 25 | 25 | 25 |

**Table 4.3.3** Output from SPSS for a chi-squared test

| Test statistics | VAR00001 |
| --- | --- |
| Chi-square[a] | 23.200 |
| Degrees of freedom | 3 |
| Asymp. Sig. | 0.000 |

[a]No cells (0.0%) have expected frequencies less than 5. The minimum expected cell frequency is 25.0.

We can then calculate the chi-squared value for this. Table 4.3.3 provides an example of the output from a chi-squared test carried out on SPSS. What we are interested in here are the two bottom outputs. In the past this would have been used to look up our calculated chi-square value against a threshold value in a set of statistical tables, however, computer packages now make this unnecessary. The significance value ('Aysmp. Sig.') at the bottom of the table is the most important output as it tells us whether there is a significant difference between our observed and expected values. If it falls below our defined significance level (normally 5% or 0.05) then we have a significant result. In this case we do have a significant difference – i.e. the number is $< 0.05$ (Table 4.3.3). Furthermore, as DF exceeds zero, the test is valid.

Chi-squared can be very useful but a drawback is that it is not appropriate for very small sample sizes. As a general rule, if there are less than 20 data values or the expected value is less than 5 (5%), then the test cannot be applied. This is why it is important to know what test you are planning to use at the experimental planning stage in order to avoid these potential problems (see section 2.2).

If we want to determine whether there is a significant difference between more than two groups then we could carry out a whole series of individual comparisons between each group – e.g. t-tests comparing dataset A with dataset B, then dataset A with dataset C, and dataset B with dataset C. However, this would not only be time consuming but probability would likely result in us finding a significant difference between some of the groups even if there were none in the context of the entire dataset comprising all groups. Therefore we can use tests that are able to test for differences between several groups at once. The best way of doing this is to use a **one-way** ANOVA test that is designed specifically for this purpose. This test examines the variance that occurs within the data and determines whether variance between groups is greater than that found within the data as a whole. While it is the parametric version of this test that is most commonly used, a non-parametric version does exist.

We can take an example where we wish to establish whether there are significant differences in macroinvertebrate density in six rivers within the same catchment. Do not forget that there needs to be replication in order to carry out statistical analysis: in this case we have five replicate samples from each site (Table 4.3.4).

Input these data into a suitable statistics package. Before we use a one-way ANOVA to determine whether the site (river) significantly affects macroinvertebrate density

**Table 4.3.4** Macroinvertebrate density (number per kick sample) in five replicate samples taken in six different rivers

|  | | River | | | |
|---|---|---|---|---|---|
| 1 | 2 | 3 | 4 | 5 | 6 |
| 152 | 112 | 65 | 176 | 98 | 45 |
| 143 | 132 | 76 | 180 | 102 | 47 |
| 100 | 130 | 86 | 198 | 110 | 50 |
| 125 | 118 | 66 | 195 | 106 | 52 |
| 136 | 121 | 70 | 187 | 113 | 60 |

we need to consider one other assumption of the test. The variability of the data in each treatment is assumed to be similar (i.e. variances are homogeneous). We can check this easily by testing the data for variance (e.g. this can be done as part of the ANOVA test in SPSS). Table 4.3.5 shows the output from SPSS 15.0 for this test.

The important values we need to look at are the F value, degrees of freedom (DF) and the $p$-value (denoted in Table 4.3.5 as Sig.). The F value relates to the size and consistency of the differences between the data from the different groups. Higher F values indicate clearer differences between means. Remember from before that if we are only asking one question (i.e. do invertebrate densities vary with site) then the degrees of freedom are simply $n - 1$, which in this case is 29, denoting that the test is valid. The $p$-value tells us whether the differences between the groups occurred by chance. The $p$-value in this case is $< 0.05$ which tells us that the likelihood of obtaining an F value of 97.646 is less than 0.05 and therefore there is a difference between at least some of the rivers. It is important to note that the $p$-value here is given as 0.000 but this simply means that the value is $< 0.001$ and the software does not display any more decimal places. Also provided is the Levene statistic, which is used to determine the homogeneity of variance. The $p$-value (Sig.) is $> 0.05$, therefore the variances are homogeneous and meet the assumptions of the ANOVA test.

Although this output will tell us whether there is a difference between the groups it does not identify which group is different from which. This requires another test called *post-hoc* testing. There are a whole raft of different tests available and each has specific assumptions associated with it and detailed descriptions of all these can be

**Table 4.3.5** Output from SPSS for one-way ANOVA: density versus site

|  | Sum of squares | DF | Mean square | F | Sig. |
|---|---|---|---|---|---|
| Between groups | 57358.700 | 5 | 11471.740 | 97.646 | 0.000 |
| Within groups | 2819.600 | 24 | 117.483 | | |
| Total | 60178.300 | 29 | | | |
| Levene statistic | DF1 | DF2 | Sig. | | |
| 2.350 | 5 | 24 | 0.072 | | |

found in Zar (1998). Unfortunately all of these require parametric data (although remember that in some cases you can transform your data to render it amenable to such tests – see section 3.3). These tests can be carried out on most statistical packages as part of the one-way ANOVA. For example, in the case above we use a Tukey test to determine where the differences between the sites lie. The output from SPSS 15.0 is shown in Table 4.3.6.

**Table 4.3.6** Output from SPSS 15.0 showing the results of a post-hoc Tukey test identifying differences in macroinvertebrate density between six different rivers

**Multiple comparisons**

Dependent variable: VAR00001
Tukey HSD

| (I) VAR00002 | (J) VAR00002 | Mean difference (I–J) | Std. Error | Sig. | 95% Confidence interval | |
|---|---|---|---|---|---|---|
| | | | | | Lower bound | Upper bound |
| 1.00 | 2.00 | 8.60000 | 6.85517 | .806 | −12.5957 | 29.7957 |
| | 3.00 | 58.60000* | 6.85517 | .000 | 37.4043 | 79.7957 |
| | 4.00 | −56.00000* | 6.85517 | .000 | −77.1957 | −34.8043 |
| | 5.00 | 25.40000* | 6.85517 | .012 | 4.2043 | 46.5957 |
| | 6.00 | 80.40000* | 6.85517 | .000 | 59.2043 | 101.5957 |
| 2.00 | 1.00 | −8.60000 | 6.85517 | .806 | −29.7957 | 12.5957 |
| | 3.00 | 50.00000* | 6.85517 | .000 | 28.8043 | 71.1957 |
| | 4.00 | −64.60000* | 6.85517 | .000 | −85.7957 | −43.4043 |
| | 5.00 | 16.80000 | 6.85517 | .179 | −4.3957 | 37.9957 |
| | 6.00 | 71.80000* | 6.85517 | .000 | 50.6043 | 92.9957 |
| 3.00 | 1.00 | −58.60000* | 6.85517 | .000 | −79.7957 | −37.4043 |
| | 2.00 | −50.00000* | 6.85517 | .000 | −71.1957 | −28.8043 |
| | 4.00 | −114.60000* | 6.85517 | .000 | −135.7957 | −93.4043 |
| | 5.00 | −33.20000* | 6.85517 | .001 | −54.3957 | −12.0043 |
| | 6.00 | 21.80000* | 6.85517 | .041 | .6043 | 42.9957 |
| 4.00 | 1.00 | 56.00000* | 6.85517 | .000 | 34.8043 | 77.1957 |
| | 2.00 | 64.60000* | 6.85517 | .000 | 43.4043 | 85.7957 |
| | 3.00 | 114.60000* | 6.85517 | .000 | 93.4043 | 135.7957 |
| | 5.00 | 81.40000* | 6.85517 | .000 | 60.2043 | 102.5957 |
| | 6.00 | 136.40000* | 6.85517 | .000 | 115.2043 | 157.5957 |
| 5.00 | 1.00 | −25.40000* | 6.85517 | .012 | −46.5957 | −4.2043 |
| | 2.00 | −16.80000 | 6.85517 | .179 | −37.9957 | 4.3957 |
| | 3.00 | 33.20000* | 6.85517 | .001 | 12.0043 | 54.3957 |
| | 4.00 | −81.40000* | 6.85517 | .000 | −102.5957 | −60.2043 |
| | 6.00 | 55.00000* | 6.85517 | .000 | 33.8043 | 76.1957 |
| 6.00 | 1.00 | −80.40000* | 6.85517 | .000 | −101.5957 | −59.2043 |
| | 2.00 | −71.80000* | 6.85517 | .000 | −92.9957 | −50.6043 |
| | 3.00 | −21.80000* | 6.85517 | .041 | −42.9957 | −.6043 |
| | 4.00 | −136.40000* | 6.85517 | .000 | −157.5957 | −115.2043 |
| | 5.00 | −55.00000* | 6.85517 | .000 | −76.1957 | −33.8043 |

*. The mean difference is significant at the .05 level.

| Site | 6 | 3 | 5 | 2 | 1 | 4 |
|------|------|------|-------|-------|-------|-------|
| Mean | 50.8 | 72.6 | 105.8 | 122.6 | 131.2 | 187.2 |

**Figure 4.3.1** Identification of differences between groups following a Tukey test.

The table shows whether macroinvertebrate densities in each river differ significantly or not from those in other rivers as indicated by the column 'sig.'. Remember as always we are looking for values < 0.05, which would indicate a significant difference. So in the first block of five rows we can see that river 1 differs significantly from rivers 3, 4, 5 and 6 (the sig. value for all < 0.05); but not river 2 (for which sig > 0.05 at 0.806). You can summarise these differences by putting the different groups in rank order as seen in Figure 4.3.1. The lines indicate no significant difference between the relevant groups, therefore rivers 2 and 5, and rivers 1 and 2 have similar macroinvertebrate densities. Rivers 3, 4 and 6 differ from all other sites.

The example used above is quite simple in that we only wanted to determine whether there was a difference in population density between the rivers – i.e. there is a single level of grouping. However, in many situations we may be interested in two factors (two levels of grouping) so we may want to determine not only whether density differs between the rivers but also whether this difference changes with season. The sample data can be found in Table 4.3.7. In this case we can use a **two-way** ANOVA, which not only determines whether there is a difference between the rivers and a difference between seasons for any given river, but also whether the differences between the rivers is significantly different between the seasons (the interaction of the two factors). The test works in a similar way to the one-way ANOVA and a similar output is obtained, but this time we also have results for the interaction (river and season).

As we can see from Table 4.3.8 not only is there a significant difference between the rivers, but also between season (i.e. in both cases $p < 0.001$). Most importantly though, it shows that the differences between the rivers also varies with season (the interaction) (i.e. for the row 'River*Season', $p < 0.001$). The main drawback with a two-way ANOVA is that as with the one-way ANOVA, it only tells us whether there is a significant difference, but not where it lies (i.e. we do not know which river has a higher density and in which season). Unfortunately the standard statistics packages at present cannot automatically apply the post-hoc tests for the two-way ANOVA and therefore this must be carried out by hand. However, this is relatively simple.

Let us take a simple example first. Say we had conducted an experiment to determine the effect of pH and the presence of cadmium on the biomass of a plant species (Table 4.3.9). We need to carry out the two-way ANOVA first, which gives the following result (Table 4.3.10). The results in Table 4.3.10 confirm that there is a significant effect of pH, cadmium and their interaction on the biomass of

**Table 4.3.7** Macroinvertebrate density in five replicate samples taken in six different rivers on four occasions during a year

| River | Sample | Season | | | |
|---|---|---|---|---|---|
| | | 1 | 2 | 3 | 4 |
| 1 | A | 152 | 165 | 150 | 45 |
| | B | 143 | 154 | 148 | 34 |
| | C | 100 | 176 | 121 | 66 |
| | D | 125 | 180 | 101 | 34 |
| | E | 136 | 178 | 98 | 57 |
| 2 | A | 112 | 127 | 86 | 50 |
| | B | 132 | 128 | 87 | 54 |
| | C | 130 | 131 | 99 | 44 |
| | D | 118 | 133 | 67 | 35 |
| | E | 121 | 115 | 78 | 46 |
| 3 | A | 65 | 89 | 45 | 23 |
| | B | 76 | 90 | 44 | 17 |
| | C | 86 | 69 | 46 | 21 |
| | D | 66 | 78 | 50 | 18 |
| | E | 70 | 89 | 53 | 25 |
| 4 | A | 176 | 200 | 145 | 112 |
| | B | 180 | 198 | 176 | 120 |
| | C | 198 | 178 | 160 | 132 |
| | D | 195 | 168 | 154 | 114 |
| | E | 187 | 197 | 143 | 120 |
| 5 | A | 98 | 124 | 100 | 100 |
| | B | 102 | 118 | 112 | 80 |
| | C | 110 | 128 | 101 | 78 |
| | D | 106 | 120 | 99 | 88 |
| | E | 113 | 122 | 97 | 90 |
| 6 | A | 45 | 66 | 46 | 10 |
| | B | 47 | 56 | 38 | 8 |
| | C | 50 | 60 | 40 | 11 |
| | D | 52 | 78 | 33 | 15 |
| | E | 60 | 59 | 36 | 13 |

**Table 4.3.8** Output from two-way ANOVA conducted using Minitab 14.0 for macroinvertebrate density measured in six rivers ($n = 5$) (Table 4.3.7)

| Source | DF | SS* | MS** | F | $p$-value |
|---|---|---|---|---|---|
| River | 5 | 192 239 | 38 443 | 345.38 | 0.000 |
| Season | 3 | 84 184 | 28 061 | 252.08 | 0.000 |
| River and season | 15 | 15 373 | 1025 | 9.21 | 0.000 |
| Error | 96 | 10 687 | 111 | | |
| Total | 119 | 302 483 | | | |

*SS = Sum of squares
**MS = Mean squares

**Table 4.3.9**  Biomass production of *Typha latifolia* (dry weight in g) grown in two different pH conditions, with and without the presence of cadmium ($n = 5$)

|  | Sample | pH | |
|---|---|---|---|
|  |  | 3.0 | 6.5 |
| Without cadmium | 1 | 4.0 | 5.4 |
|  | 2 | 4.2 | 5.6 |
|  | 3 | 4.6 | 5.5 |
|  | 4 | 4.0 | 5.3 |
|  | 5 | 4.5 | 5.5 |
| With cadmium | 1 | 1.3 | 4.3 |
|  | 2 | 1.5 | 4.5 |
|  | 3 | 1.4 | 4.9 |
|  | 4 | 1.1 | 4.5 |
|  | 5 | 1.4 | 4.3 |

*Typha latifolia* (i.e. *p*-values all $< 0.001$) but we do not know what the effects are exactly. To establish this we need to conduct a post-hoc test, in this case a Tukey test.

The first step is to put the mean values for each group into ascending numerical order as we did for the one-way ANOVA (in Minitab one of the outputs from the ANOVA is a list of the mean values). So for our example above we obtain (Table 4.3.11):

From results it looks fairly obvious that plants grown in the lower pH (3.0) medium in the presence of cadmium have much lower biomass than the other plants. However, in larger datasets this may not be so obvious, so we need a further step to establish this. We need to determine two values, and the first of these is the **standard error** (SE; see section 3.1), which can be calculated from the values generated by the ANOVA.

$$SE = \sqrt{\frac{ErrorMS}{n}}$$

**Table 4.3.10**  Output from Minitab 14 for two-way ANOVA to determine the effect of pH and cadmium, and their interaction on biomass production of *Typha latifolia*

| Source | DF | SS* | MS** | F | *p*-value |
|---|---|---|---|---|---|
| pH | 1 | 23.762 | 23.762 | 546.25 | 0.000 |
| Cadmium | 1 | 18.818 | 18.818 | 432.6 | 0.000 |
| pH and cadmium | 1 | 4.802 | 4.802 | 110.39 | 0.000 |
| Error | 16 | 0.696 | 0.044 | | |
| Total | 19 | 48.078 | | | |

*SS = Sum of squares
**MS = Mean squares

**Table 4.3.11**  Mean values of biomass in numerical order

|             | 1       | 2       | 3       | 4       |
|-------------|---------|---------|---------|---------|
| pH          | 3.0     | 3.0     | 6.5     | 6.5     |
| Cadmium     | With    | Without | With    | Without |
| Mean biomass| 1.34    | 4.26    | 4.5     | 5.46    |

The *ErrorMS* can be found in the output from the ANOVA (Table 4.3.10) and $n$ is simply the number of replicates (in this case 5).

$$SE = \sqrt{\frac{0.044}{5}}$$

and hence SE = 0.094.

Now we need to calculate a $q$-value for each pair of means in sequence. This value simply tells us the difference between the means divided by the SE (which we have just calculated). So for the first two mean values

$$4.26 - 1.34 = 2.92$$
$$q = 2.92/0.094$$
$$q = 31.06$$

Now we need to establish whether this figure is significant or not, and we do that by comparing it to a critical $q$-value which can be found in statistical tables (Lindley and Scott, 1995). The critical $q$-value is found in the tables for $p = 0.05$ and the error degrees of freedom ($v$) (found in the ANOVA output – Table 4.3.10) and the number of means being compared ($k$). So in this case the error degrees of freedom are 16, the number of means are 4, and the critical $q$ is thus 3.007.

We then compare the critical $q$ to our calculated $q$ from before and if the calculated $q$ is greater than the critical $q$ from the tables then the difference is significant. Our calculated $q$ was 31.06, which is greater than 3.007 and therefore the difference is significant (i.e. plants grown in the presence of cadmium at pH 3 have significantly lower biomass than plants grown at the same pH without cadmium).

We can tabulate this to make it clearer (Table 4.3.12). Note that because 1 was significantly different from 2 and we have put them in ascending order, then 1 must

**Table 4.3.12**  Outcome of post-hoc Tukey test on plant biomass data

| Comparison | Difference | SE | Calculated $q$ | Critical $q$ | Outcome |
|------------|-----------|-------|-------------|-----------|-----------------|
| 1 versus 2 | 2.92 | 0.094 | 31.06 | 3.007 | Significant |
| 2 versus 3 | 0.24 | 0.094 | 2.55 | 3.007 | Not significant |
| 2 versus 4 | 1.20 | 0.094 | 12.77 | 3.007 | Significant |
| 3 versus 4 | 0.96 | 0.094 | 10.43 | 3.007 | Significant |

| 1 | 2 | 3 | 4 |

**Figure 4.3.2** Identification of differences between groups following calculation of the $q$ value.

also be significantly different from 3 and therefore there is no need to calculate the $q$ value. If we find that two means are not significant (as for 2 versus 3) then move on to compare the lowest value with the next value (2 versus 4), and so on. We can then illustrate the results as we did for the one-way ANOVA (Figure 4.3.2). Now we can say that there is no significant difference in biomass production of plants grown with cadmium at pH 6.5 and those grown without cadmium at pH 3.0.

Because in this specific example we have only two levels in any treatment it is clear that any differences found to be significant must be between those two levels. In some experiments, however, there may be more than one level in any treatment. So if we return to our river macroinvertebrate density data (Table 4.3.7), we can see that there are four levels within the season 'treatment' and six levels within the 'river' treatment. This does not mean that we cannot identify the differences using a Tukey (or other post-hoc) test, it simply means that we have to make multiple comparisons. This is achieved in exactly the same way as we have just done but when calculating SE the $n$ value is the number of data values at each level of treatment being tested. So if we were examining the effect of the river then $n$ would be 20 (for each river we have five replicate samples from four seasons therefore it is $5 \times 4 = 20$). This value is provided for each treatment and the interaction, in the output from the statistical tests (Tables 4.3.13 and 4.3.14). We can see that there is a significant difference between all the rivers, which you might not have expected by simple inspection of the data. We can now repeat this for both the effect of season and then for the interaction between season and river. For the latter you will end up with a large table and many comparisons, which increases the complexity of interpreting the results.

**Table 4.3.13** Mean values of macroinvertebrate density in ascending numerical order

| Ranking | 1 | 2 | 3 | 4 | 5 | 6 |
|---|---|---|---|---|---|---|
| River | 6 | 3 | 2 | 5 | 1 | 4 |
| Mean density | 41.15 | 56.0 | 94.65 | 104.30 | 118.15 | 162.65 |

**Table 4.3.14** Outcome of post-hoc Tukey test on macroinvertebrate data

| Comparison | Difference | SE | Calculated $q$ | Critical $q$ | Outcome |
|---|---|---|---|---|---|
| 1 versus 2 | 14.85 | 3.33 | 4.45 | 2.191 | Significant |
| 2 versus 3 | 38.65 | 3.33 | 11.61 | 2.191 | Significant |
| 3 versus 4 | 9.65 | 3.33 | 2.91 | 2.191 | Significant |
| 4 versus 5 | 14.2 | 3.33 | 4.26 | 2.191 | Significant |
| 5 versus 6 | 44.5 | 3.33 | 13.36 | 2.191 | Significant |

**Table 4.3.15** Illustrative output from two-way ANOVA for interaction between river and season for data in Table 4.3.7

| Source | DF | SS* | MS** | F | p-value |
|---|---|---|---|---|---|
| River | 5 | 192239 | 38448 | 345.38 | 0.000 |
| Season | 3 | 84184 | 28061 | 252.08 | 0.000 |
| River and season | 15 | 15373 | 1025 | 9.21 | 0.000 |
| Error | 96 | 10687 | 111 | | |
| Total | 119 | 302483 | | | |

*SS = Sum of squares
**MS = Mean squares

Use the data in Table 4.3.7 and carry out a two-way ANOVA and the post hoc Tukey test for river, season and their interaction for yourself. You should find that from the two-way ANOVA the results shown in Table 4.3.15 are obtained.

The Tukey test in this case is quite complicated and the easiest way to display the outcome is using a graph or table (Figure 4.3.3). You can then indicate where the differences lie using some sort of code (in this case we have used letters). More information on how to present data can be found in Chapter 7.3.

There is no reason why we have to stop at two treatments (i.e. river and season), and we can have three, four or even more treatments. However, the more you have the more difficult it is to interpret the data and the more likely it is that the data are no longer independent (you can see how difficult it is to interpret just two

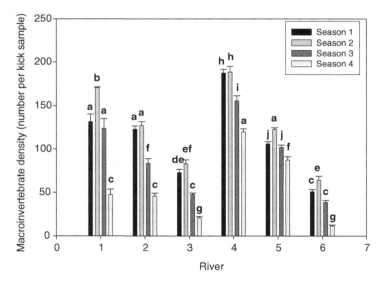

**Figure 4.3.3** Macroinvertebrate densities (number per 3 minute kick sample) in six different rivers according to season. Bars indicate standard error ($n = 5$). Different letters indicate a significant difference (two-way ANOVA, $p < 0.05$).

treatments from the data above; Figure 4.3.3). Hence, more than three levels are usually avoided.

The two-way ANOVA is a very useful test for many experimental designs, however, if you have unbalanced data (i.e. missing values) or unequal replication then the test cannot be applied. An alternative approach that can be used is the General Linear Model, which can cope with unbalanced data (Rutherford, 1997).

## 4.3.2 Non-parametric tests

Earlier we discussed alternative strategies for dealing with data that are not normally distributed. Remember that we can transform the data (see section 3.3), but in some cases this may not solve the problem. In such instances, we cannot use parametric tests, but instead need to use non-parametric tests.

For parametric data where we want to determine whether there is a significant difference between two groups then we use either a paired or an independent t-test. Where we have paired samples and non-parametric data, instead of applying the paired t-test we can use the **Wilcoxon signed rank test**. The reason why it can be applied to non-parametric data is that the test is based on the rank order of the differences between the two groups rather than the actual value of the differences. However, the assumption is still made that the distribution of the differences is symmetric. It should be noted that non-parametric tests can be applied to parametric data (but not vice versa), so if for some reason you are not sure as to the normality of your data it is best to apply the non-parametric tests (but remember they are less powerful).

As an example, we may want to determine whether a water treatment system removing iron from effluent is working and therefore we take samples at the inlet and outlet. We would expect the two to be related as we take the samples at the same time and therefore the samples are paired. We can use a Wilcoxon signed rank test to establish whether there is a significant difference between concentrations at the outlet and inlet. Input the data in Table 4.3.16 into a statistical package (e.g. Minitab or SPSS) and carry out this test, which will be listed under non-parametric tests. You should find that $p < 0.001$ and therefore there is a significant difference between inlet and outlet concentrations.

Where we have independent samples (e.g. when we are comparing outlet concentrations of iron at two different treatment systems), an alternative to the parametric independent t-test is the **Mann–Whitney U-test**. This test is able to deal not only

**Table 4.3.16** Iron concentrations (mg $L^{-1}$) in inlet and outlet waters within a treatment system ($n = 20$)

| Inlet | 20 | 24 | 23 | 21 | 20 | 29 | 27 | 20 | 24 | 25 | 20 | 20 | 19 | 21 | 23 | 19 | 24 | 23 | 24 | 25 |
|-------|----|----|----|----|----|----|----|----|----|----|----|----|----|----|----|----|----|----|----|----|
| Outlet | 8 | 6 | 3 | 4 | 5 | 10 | 4 | 3 | 5 | 10 | 9 | 8 | 2 | 5 | 6 | 6 | 5 | 10 | 8 | 6 |

with raw counts but also ordinal and constant interval data, and as for the independent t-test can cope with unequal sample sizes in the two groups. As with the Wilcoxon test the Mann–Whitney U-test is designed specifically for non-parametric data but it is assumed that the two groups have a similar distribution, i.e. if one group is negatively skewed, then the other group must also be negatively skewed (see section 3.2 and Figure 3.2.4B). If the two sets of data have very different distributions then they cannot be analysed this way. It should also be emphasised that like t-tests, this test is only applicable to two groups (not more than two groups) and can only be used to test a two-tailed prediction.

Carry out a Mann–Whitney U-test on the data in Table 4.3.1 (remember non-parametric tests can be used on parametric data). You should find that there is a significant difference between the two groups as we found previously using the parametric t-test.

The non-parametric alternative to the one-way ANOVA is the Kruskal–Wallis test. Again, although this test can cope with data that are not normally distributed, it still assumes that the variances of the two groups are equal and the distributions the same. An example of where a Kruskal–Wallis test would be applied is if we wanted to determine whether there was a significant difference between the number of eggs laid by a particular moth on different plant species. The results of this experiment can be found in Table 4.3.17.

If we carry out a Kruskal–Wallis test to establish whether there is a significant difference in egg counts between the different species then we get the following results (Table 4.3.18). As for the one-way ANOVA we are given the degrees of freedom (DF) and a $p$-value of $< 0.001$, indicating a significant difference between the plant species. In addition we are given the H-value, which is equivalent to the t-value in     t-tests.

The z-values give us an indication of where the differences lie. The smaller the z-value the less different the average rank for the treatment is from the average rank for all observations. So we can see that it is plant species 8 that is significantly higher than the mean rank as the value is positive and plant species 3 that is significantly

**Table 4.3.17** Moth egg counts on ten different plant species ($n = 10$)

| Plant species | Egg count |
|---|---|
| 1 | 10, 15, 20, 18, 60, 25, 20, 10, 15, 14 |
| 2 | 32, 68, 70, 15, 13, 20, 21, 19, 24, 15 |
| 3 | 16, 13, 14, 34, 12, 15, 18, 19, 21, 22 |
| 4 | 14, 15, 21, 27, 66, 12, 15, 16, 19, 29 |
| 5 | 24, 21, 24, 23, 13, 16, 12, 15, 24, 51 |
| 6 | 52, 12, 13, 18, 95, 100, 105, 114, 23, 24 |
| 7 | 34, 36, 35, 46, 35, 38, 42, 43, 45, 38 |
| 8 | 22, 36, 37, 44, 58, 104, 43, 48, 104, 24 |
| 9 | 25, 35, 39, 47, 24, 38, 99, 104, 35, 32 |
| 10 | 31, 22, 41, 40, 30, 34, 116, 34, 105, 38 |

**Table 4.3.18** Output from Minitab 14.0 for a Kruskal–Wallis test: plant species against number of moth eggs

| Species | $n$ | Median | Average rank | $z$ |
|---|---|---|---|---|
| 1 | 10 | 16.5 | 27.8 | −2.61 |
| 2 | 10 | 20.5 | 42.2 | −0.96 |
| 3 | 10 | 17 | 25.5 | −2.88 |
| 4 | 10 | 17.5 | 33.2 | −1.99 |
| 5 | 10 | 22 | 35.3 | −1.75 |
| 6 | 10 | 38 | 59.2 | 1.00 |
| 7 | 10 | 38 | 70.9 | 2.34 |
| 8 | 10 | 43.5 | 73.5 | 2.64 |
| 9 | 10 | 36.5 | 69.3 | 2.15 |
| 10 | 10 | 36 | 68.4 | 2.06 |
| H = 40.84: DF = 9: $p = 0.000$ | | | | |
| H = 40.90: DF = 9: $p = 0.000$ (adjusted for ties) | | | | |

lower as the value is negative. Plant species 2 is least different from the mean rank. Unfortunately this is as far as we can go using the overall Kruskal–Wallis test and unlike the one-way ANOVA we cannot carry out post-hoc tests to determine where the differences lie. It is possible to carry out Mann-Whitney U-tests to compare each sample against the others in turn but this is not ideal as results often differ from the output from the Kruskal–Wallis test.

# References

Lindley, D.V. and Scott, W.F. 1995. *New Cambridge Statistical Tables*, Cambridge University Press, Cambridge.

Rutherford, A. 1997. *Introducing ANOVA and ANCOVA: A GLM Approach*, Sage, London.

Zar, J.H. 1998. *Biostatistical Analysis*, Prentice Hall, Harlow.

# 5

# Spotting relationships

## 5.1 Linear regression – to what extent does one factor influence another?

One of the most common types of question posed in environmental science are those of the form 'if variable $x$ is altered, will there be a proportional change in variable $y$?'. Related to this type of question, it is usually useful to know the type and magnitude of change in $y$ brought about by a change in $x$. For example, if we increase variable $x$ by a factor of 10%, will it produce a change in variable $y$, and if so, will the change in $y$ be an increase or a decrease, and by what percentage?

Examples of this type of question are:

- If we increase the concentration of a pollutant in a river (i.e. variable $x$), will there be an increase in deformities in Chironimidae (midge larvae) (variable $y$)? If there is an increase, how large will it be?

- If we reduce the number of vehicles on a road (i.e. variable $x$), will be there be a decrease in air pollution (variable $y$)? If there is a decrease, how large will it be?

To answer such questions about two possibly related variables, one can utilise a technique known as linear regression. The technique involves plotting the values of variable $y$ (as the $y$-axis) obtained at the corresponding values of $x$ (plotted as the $x$-axis). This is far easier and indeed far better conducted using a software package such as Excel, SPSS or Minitab. For a detailed treatment of how to use Excel to conduct linear regression, you are referred to the excellent text by Keller (2001).

When using linear regression in environmental science, it is important to adhere to the following principle. In the examples above (and any others that you might care to think of), we are examining whether, and to what extent, the value of the $y$ variable **depends** on a change in the value of the $x$ variable. It is for this reason that the

*Student Projects in Environmental Science*   Edited by Stuart Harrad and Lesley Batty
© 2008 John Wiley & Sons Ltd

**Table 5.1.1**  Data used to plot Figures 5.1.1 and 5.1.2

| x | y |
|---|---|
| 1.2 | 7.4 |
| 3.4 | 11.8 |
| 5.6 | 16.2 |
| 7.1 | 19.2 |
| 11.5 | 28 |
| 23.7 | 52.4 |
| 29 | 63 |
| 32.1 | 69.2 |
| 37.8 | 80.6 |
| 45 | 95 |
| 56.9 | 118.8 |
| 73.4 | 151.8 |
| 89.2 | 183.4 |

$y$ variable is referred to as the **dependent** variable, and the $x$ variable as the **independent** variable (see section 2.2). In the first example above, we are trying to find out the extent to which the incidence in deformities of Chironimidae ($y$ variable) depends on the value of pollutant concentration ($x$ variable).

Table 5.1.1 contains a set of values for two variables $x$ and $y$. When plotted against each other in Excel, one obtains a graph similar to that shown in Figure 5.1.1. Looking at this graph, you should be able to see that as $x$ increases, so does $y$, so just by simple visual inspection, there appears to be a relationship between the two variables $y$ and $x$. What linear regression does is provide information that enables us to assess:

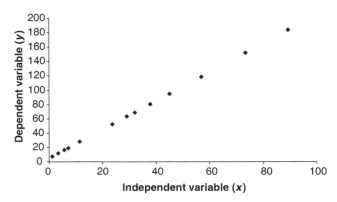

**Figure 5.1.1**  Plot of dependent variable (y) versus independent variable (x) (data from Table 5.1.1).

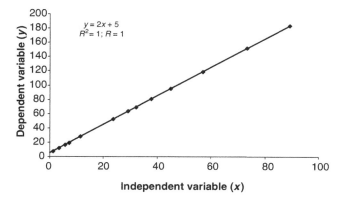

**Figure 5.1.2** Plot of dependent variable (y) versus independent variable (x) with information relating to linear relationship between y and x added (from Table 5.1.1).

- whether a linear (i.e. a straight line) relationship exists;

- what the relationship is (e.g. $y = 2x + 5$) – note here that the slope (or gradient) of the straight line $= 2$, and the $y$-intercept (i.e. the value of $y$ when $x = 0$) $= 5$;

- how significant (i.e. how reliable) the relationship is.

In Figure 5.1.2 you will see we have used Excel to draw a straight line through the data points. Indeed, the straight line runs exactly through all the points, and we can say that the relationship is perfectly linear, and that the two variables are correlated perfectly. Excel has also given us a measure of how well linearly correlated (or how closely the data points adhere to the straight line) the two variables are. This measure is either the Pearson product moment correlation coefficient (hereinafter referred to as the correlation coefficient – $R$) or its square ($R^2$). In Figure 5.1.2, the variables are correlated perfectly and both $R$ and $R^2 = 1$. Essentially, the stronger the linear correlation between two variables, the closer the value of $R$ (and thus $R^2$) will be to 1.

Table 5.1.2 gives example data for how airborne concentrations of two different chemical pollutants known as PCB#18 and PCB#101, depend on the reciprocal of temperature (i.e. 1/temperature). Here the $x$ or independent variable is the reciprocal temperature, and the concentrations (expressed in the probably unfamiliar units of Ln Pa) are the dependent or $y$ variables. When these are plotted against each other, Figures 5.1.3 and 5.1.4 result.

The data points lie fairly close to the straight line for both figures, so by eye it is not possible to detect easily which linear relationship is stronger. However, examination of the $R$ (or $R^2$) values for the two figures shows that the linear relationship is stronger (i.e. $R$ value is closer to 1) for the data shown in Figure 5.1.4. This tells us that the linear relationship between reciprocal temperature and the concentration of PCB #101 is stronger than that for PCB #18. You may be wondering at this point why both $R$ and $R^2$ are used. First, $R$ can be either positive or negative, thereby giving us

**Table 5.1.2** Data used to plot Figures 5.1.3 and 5.1.4

| Concentration[a] of PCB#18 | Concentration[a] of PCB#101 | Reciprocal temperature $(1/T)$ $(K^{-1})$ |
|---|---|---|
| −33.5 | −34.4 | 0.003448 |
| −31.9 | −33.2 | 0.003427 |
| −33.3 | −34.1 | 0.003492 |
| −33.7 | −34.8 | 0.003472 |
| −34.4 | −35.6 | 0.003646 |
| −32.8 | −34.8 | 0.003542 |
| −34.5 | −35.8 | 0.003600 |
| −34.7 | −36.1 | 0.003579 |
| −34.5 | −34.8 | 0.003542 |
| −33.2 | −34.1 | 0.003519 |
| −33.8 | −35.1 | 0.003542 |
| −33.5 | −34.9 | 0.003466 |
| −33.7 | −34.5 | 0.003483 |
| −33.1 | −33.4 | 0.003461 |
| −33.4 | −34.4 | 0.003457 |
| −33.8 | −34.5 | 0.003432 |
| −33.5 | −34.8 | 0.003477 |
| −34.1 | −35.1 | 0.003489 |
| −33.7 | −34.5 | 0.003471 |
| −34.9 | −36/0 | 0.003530 |
| −34.4 | −34.9 | 0.003595 |
| −34.9 | −35.7 | 0.003574 |
| −34.9 | −35.8 | 0.003511 |
| −34.5 | −35.1 | 0.003529 |
| −34.3 | −34.8 | 0.003539 |
| −34.0 | −34.5 | 0.003550 |
| −33.4 | −34.9 | 0.003498 |
| −33.5 | −34.5 | 0.003487 |
| −33.8 | −35.0 | 0.003537 |
| −33.0 | −34.0 | 0.003436 |
| −33.6 | −34.6 | 0.003510 |
| −33.9 | −35.6 | 0.003592 |
| −34.7 | −35.8 | 0.003566 |
| −33.3 | −35.9 | 0.003662 |
| −34.5 | −34.9 | 0.003623 |
| −35.3 | −35.7 | 0.003610 |
| −33.5 | −33.4 | 0.003565 |
| −33.7 | −35.1 | 0.003527 |
| −32.7 | −34.0 | 0.003471 |
| −32.8 | −34.3 | 0.003483 |
| −33.4 | −34.4 | 0.003544 |

[a]Expressed as the natural logarithm of the partial pressure (Ln Pa)

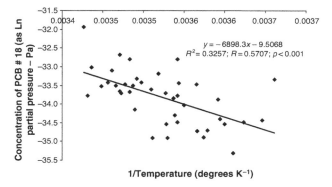

**Figure 5.1.3** Plot of relationship between concentration of PCB #18 versus reciprocal temperature (from Table 5.1.2).

an indication of whether increasing the independent variable causes an increase (i.e. $R$ is positive) or decrease (i.e. $R$ is negative) in the dependent variable, whereas $R^2$ by definition can only be positive. In both cases here, $R$ is negative. This denotes that increasing reciprocal temperature causes the pollutant concentration to decrease. Put another way, increasing temperature increases the pollutant concentration. Second, $R^2$ is an expression of the extent to which the variation in values of the dependent variable is due to variations in values of the independent variable. In Figure 5.1.3, the $R^2$ value is 0.3257, this indicates that 32.57% (i.e. $0.3257 \times 100\%$) of the variability in concentrations of PCB #18 (the dependent variable) is due to changes in the reciprocal temperature (the independent variable). In Figure 5.1.4, the $R^2$ value is

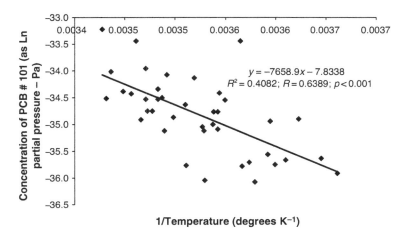

**Figure 5.1.4** Plot of relationship between concentration of PCB #101 versus reciprocal temperature (from Table 5.1.2).

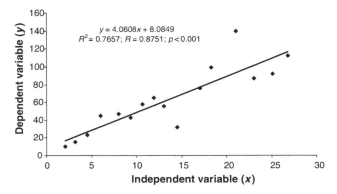

**Figure 5.1.5**  Plot of dependent variable ($y$) versus independent variable ($x$) (16 data points).

0.4082, which tells us that 40.82% of the variability in concentrations of PCB #101 is due to changes in reciprocal temperature. In other words, concentrations of PCB #101 are more temperature-dependent than those of PCB #18.

While – as shown above – simple comparison of $R$ values can tell us the relative strength of linear correlation between various sets of data for two variables, such a comparison is only strictly valid when – as in Table 5.1.2 – the number of data points in each dataset is equal. To illustrate what can happen when one simply compares values of $R$, look at Figures 5.1.5 and 5.1.6.

In Figure 5.1.5, there are 16 data points, for which the $R$ value is 0.8751. By comparison, the $R$ value for the dataset plotted in Figure 5.1.6 is slightly closer to 1 (i.e. 0.8820). However, the number of data points plotted in Figure 5.1.6 is only five – i.e. considerably less than in Figure 5.1.5, and it should be apparent that although the value for $R$ is greater for Figure 5.1.6, two out of the five data points (i.e. 40%) are some distance from the straight line. By comparison, in Figure 5.1.5, only 12.5% (i.e.

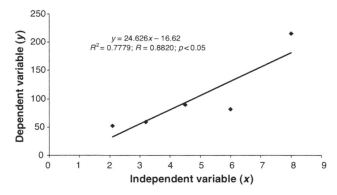

**Figure 5.1.6**  Plot of dependent variable ($y$) versus independent variable ($x$) (five data points).

two out of 16) data points are similarly distant from the straight line. On this basis, common sense indicates that the linear relationship is stronger for the larger dataset plotted in Figure 5.1.5. However, we still require a quantitative expression of what such observation and common sense tells us. This is supplied by the $p$-value. The $p$-value is an expression of the statistical significance of the linear relationship (see section 3.2). In essence, the smaller the $p$-value, the closer the data points lie to the straight line, and the stronger is the linear relationship. A $p$-value of 0.001 in this example means that we can be 99.9% (i.e. $100 \times (1 - 0.001)$) confident that a linear relationship exists between the two variables we are investigating. Put another way, such a $p$-value tells us that there is only a one in 1000 chance ($0.001 = $ one thousandth) that a linear relationship does not exist. Similarly, a $p$-value of 0.05 means that we can be 95% (i.e. $100 \times (1 - 0.05)$) confident of the existence of a linear relationship, or that there is only a one in 20 chance that a linear relationship does not exist. This concept of confidence means that the $p$-value is sometimes referred to as the confidence level. While some software packages (e.g. Minitab, SPSS and the most recent versions of Excel – the latter is accessed via the Tools pull-down followed by Data Analysis and Regression – note that you may have to install the Data Analysis Toolpak Add-In from your Excel or Office CD) will calculate the $p$-value for you, it is possible that your software package will not. In such instances, one has to make reference to a table that gives the values of the correlation coefficient that equate to different $p$-values for different numbers of data points. Such a table may be found in a relevant textbook, e.g. Lindley and Scott (1995). On inspection of such a table you will see that rather than being based on numbers of samples, it is based on a parameter known as degrees of freedom (DF) (see section 4.3.1). In this case, DF may be easily calculated from the number of samples ($n$) as $DF = n - 2$. Remember that the test is only valid provided that $DF > 0$. This is illustrated neatly here, as for $DF = 0$, the number of data points ($n$) must $= 2$. Clearly, any two data points can always have a straight line interpolated between them, and any test that attempts to evaluate the existence of a straight line is therefore meaningless.

One thing to bear in mind when presenting your results (see section 7.3) is that while plotting data on a graph such as Figure 5.1.5 is useful, repeating similar graphs for a whole series of similar relationships (e.g. incidence of Chironimidae deformities versus sediment concentrations of a range of metals like Pb, Cd, Cu, Zn, As, Hg, Fe, etc.) can be a misuse of space. In such instances, it is recommended that you include one figure as an illustration of the relationship between one pair of variables, but summarise the relevant data (e.g. values of $R$ and $p$) for relationships between other pairs of variables. Table 5.1.3 illustrates this.

Note that you do not need to plot the graphs to obtain values of $R$, $R^2$, slope, and $y$-intercept. You can ask Excel to calculate these for you for a given set of data points using the appropriate function commands – e.g. CORREL (or PEARSON), RSQ, SLOPE, and INTERCEPT, respectively – or via Minitab or SPSS. One thing that plotting the graph does help with, is spotting whether the relationship is a positive (i.e. increasing the value of the $x$ variable results in increase in the value of the $y$ variable) or a negative one (i.e. increasing the value of the $x$ variable results in a

**Table 5.1.3** Summary of strength of linear relationships between concentration of metals in sediments and incidence of Chironimidae deformities

| Metal | $R$ | $p$ |
|-------|--------|--------|
| Hg | 0.7574 | <0.02 |
| Pb | 0.4815 | >0.1 |
| Cu | 0.7269 | <0.02 |
| Ni | 0.5910 | <0.1 |
| Zn | 0.9050 | <0.001 |
| Cd | 0.6419 | <0.05 |
| As | 0.1923 | >0.1 |

decrease in the value of the y variable). For example, in Figures 5.1.3 and 5.1.4 we can see that the relationship between PCB concentration and reciprocal temperature is negative. This means that as reciprocal temperature increases, the PCB concentration decreases, and written more intuitively, that as temperature increases, so does the PCB concentration – the underpinning scientific rationale being that PCBs evaporate more readily out of soil at higher temperatures. However, you can obtain the same information from the values of CORREL (or PEARSON), and SLOPE supplied by Excel. If the values of these parameters are positive, then it is a positive relationship, and vice versa. Note that the value of RSQ will always be positive, as the square of any number (regardless of whether it is positive or negative) will always be positive.

Box 5.1.1 provides three case studies illustrating how the techniques of linear regression can be used to help answer real questions in environmental science. The last of which illustrates a more advanced application of linear regression to an extremely topical environmental problem.

## 5.1.1  Examples of the use of linear regression

### Case study 1 – calibration of analytical instruments

An extremely widespread example of the use of linear regression in quantitative environmental science is in the calibration of analytical instruments. Here, we wish to relate instrument readings to a concentration of a chemical pollutant. By taking instrument readings for a range of known concentrations of a pollutant (these can be purchased and diluted as required), we can obtain data on the relationship between the two. By plotting instrument reading as the x (or independent) variable against concentration as the y (or dependent) variable, and using a software package such as Excel, Minitab, or SPSS, we can: (1) check that there is a statistically significant linear relationship between the two; and (2) obtain a regression equation allowing us to predict the pollutant concentration in a sample

**Table 1**  Variation of instrument reading with Pb concentration ($\mu g\ L^{-1}$)

| Instrument reading (units) | Pb concentration ($\mu g\ L^{-1}$) |
|---|---|
| 800 | 1 |
| 3800 | 5 |
| 16 500 | 20 |
| 42 100 | 50 |
| 79 200 | 100 |

from the instrument reading (i.e. how pollutant concentration depends on instrument reading).

To illustrate, look at the instrument readings below that were obtained for a range of known concentrations of Pb in water (Table 1). By plotting the Pb concentration as the $y$ variable, and the instrument reading as the $x$ variable, we obtain the graph (or plot) shown in Figure 1.

Clearly, the relationship between the two variables approximates very closely to a straight line – i.e. it is linear. We can derive quantitative information on how close it approximates to a perfect straight line from the values of $R$ (the correlation coefficient) and $p$. The value of 0.9994 denotes a $p$-value $<0.001$, and thus we can say that the relationship between Pb concentration and instrument reading is linear at a statistical significance level in excess of 99.9% (i.e. there is only a less than one in a thousand chance that the relationship is not linear). Another way of looking at things is that the $R^2$ value (simply the square of the $R$ value) is 0.9988 – in other words, 99.88% of the variability in instrument reading is caused by changes in concentration – the rest is most likely due to uncertainty in the method.

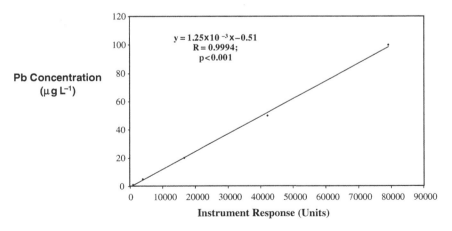

**Figure 1**  Illustrative calibration plot for Pb.

The linear relationship between concentration and instrument reading is expressed by the equation shown on the graph in Figure 1. The slope $= 1.25 \times 10^{-3}$, and the $y$-intercept value is 0.51. Try entering the values in Table 1 into your software package and experiment to obtain these values. We can subsequently use this equation to estimate Pb concentration in a water sample by substituting the instrument reading obtained for that sample into the equation. For example, the concentration of Pb in a sample giving an instrument reading of 11 900 is $(1.25 \times 10^{-3} \ 11\ 900) - 0.51 = 14.37 \mu g\ L^{-1}$.

What will be the Pb concentration in samples giving instrument readings of (1) 2350 and (2) 14 780? Remember that the Pb concentration is the $y$ or dependent variable, and the instrument reading the $x$ or the independent variable.

The establishment of the relationship between instrument reading and pollutant concentration is sometimes referred to as deriving a calibration plot, and is a corner stone of environmental analysis.

## Case study 2 – human response to noise pollution

The way in which people react or respond to an environmental stressor is extremely important, as these subjective reactions can be important when formulating policies and action plans. There are therefore occasions when it is important to be able to quantify a subjective response to an environmental stressor. This is usually achieved through the use of social surveys.

Questionnaire-based social surveys are used extensively to evaluate the impact of a noise source on an exposed population. Specifically, individuals are asked to indicate how 'annoying' a particular noise is, and to then measure or estimate the individual's noise exposure level due to that noise source. By repeating this throughout an exposed community it is then possible to produce a 'noise dose-response relationship'.

With regard to assessing the annoyance due to aircraft noise a typical question might be phrased as:

'Thinking about the past 12 months or so, when you are here at home, how much does noise from aircraft bother, disturb or annoy you? Please tick ONE box only:

Extremely
Very
Moderately
Slightly
Not at all

**Table 2**  Aircraft noise dose–response relationship

| Aircraft noise exposure level $L_{Aeq(16\,h)}$ dB | Percentage extremely annoyed |
|---|---|
| 51 | 12 |
| 54 | 22 |
| 57 | 28 |
| 60 | 36 |
| 63 | 54 |

The results derived from asking this question could then be analysed by considering the percentage of respondents who indicated that they were extremely annoyed by aircraft noise. These can then be compared with the aircraft noise exposure levels that correspond to the location of the respondent. The noise exposure level used could be obtained by the student taking noise meter measurements at residences or by analysis of a previously derived noise contour map. This is illustrated in Table 2, which uses data obtained from this question.

This can now be shown as a linear regression plot – i.e. Figure 2. From Figure 2, it is clear from the low $p$-value that there is a linear relationship between aircraft noise dose (as quantified by our questionnaire responses) and human annoyance.

It is also possible to estimate the aircraft noise exposure level at which we might expect none of the population to be extremely annoyed by aircraft noise. This can be achieved by using the regression formula shown in Figure 2, as follows.

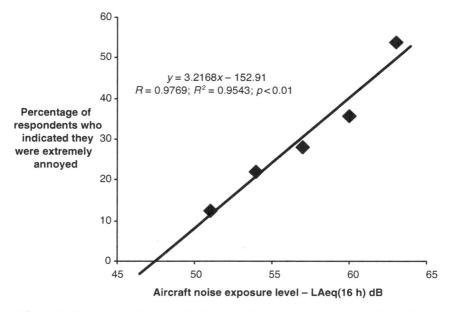

**Figure 2**  Linear regression plot to show aircraft noise dose–response relationship.

When none of the population are extremely annoyed we assume that $y = 0$, and by the following simple algebraic manipulation, we can work out the aircraft noise level that should not leave anyone extremely annoyed – i.e.

$$3.2168x - 152.91 = 0$$
$$3.2168x = 152.91$$
$$x = 152.91/3.2168 = 47.5 \, \text{dB}$$

So at an aircraft noise exposure level of 47.5 dB we would expect none of an exposed population to be extremely annoyed by aircraft noise.

As aircraft noise usually exceeds 47.5 dB, it is then pertinent to ask what percentage of an exposed population would be extremely annoyed by an aircraft noise exposure level of 65 dB? In this case the noise exposure level is 65 dB – in other words $x = 65$. So by using the regression formula given in Figure 2 we see that as

$$y = 3.2168x - 152.91$$

it follows that at 65 dB, the percentage of an exposed population that would be extremely annoyed is

$$y = (3.2168 \times 65) - 152.91$$

Hence

$$y = 209.09 - 152.91 = 56.2\%$$

So at an aircraft noise exposure level of 65 dB we would expect 56.2% of the exposed population to be extremely annoyed by aircraft noise.

# Case study 3 – what proportion of a pollutant arises from traffic emissions?

Benzo[a]pyrene (B[a]P) is an organic pollutant that is a known human carcinogen, and for which the UK Government's Expert Panel on Air Quality Standards has set an ambient air quality standard of 0.25 ng m$^{-3}$ as an annual average that has been incorporated into the UK's National Air Quality Strategy as an objective to be met by 2010. At the time of writing (2007), this standard was exceeded in some areas of the UK. A first step towards reducing concentrations to meet the standard is quantifying the relative contributions from different sources, so that action can be taken to limit emissions. There is a variety of evidence that suggests that B[a]P is emitted from traffic, and in the following example, linear regression is used to quantify the percentage contribution of traffic emissions to atmospheric concentrations of B[a]P.

We start from a well-proven assumption that in UK cities almost all (typically 98% or more) of carbon monoxide (CO) is emitted by traffic. Consequently, it is reasonable to assume that any variation in airborne concentrations of CO will be related to traffic emissions – in other words, if traffic activity is high, CO concentrations in air will be high. Therefore, if concentrations of CO and B[a]P are correlated linearly, then this suggests that traffic is an important source of B[a]P. Of course, as already mentioned, some mechanistic understanding of the processes involved is required before any causal relationship (i.e. that traffic is the source of B[a]P) can be inferred from any correlation between B[a]P and CO (see section 4.1). In this situation – given the evidence available from other sources that B[a]P can be emitted by traffic – we can be pretty sure that any correlation between B[a]P and CO is good evidence that traffic emissions are an important source of B[a]P.

To establish the existence of any correlation between concentrations of B[a]P and CO, we start by measuring both simultaneously at the same location. Owing to the difficulties in measuring B[a]P in air, a total of seven measurements of both pollutants was made at 2-h intervals between 06.00 and 20.00 hours on the same day. These are listed in Table 3.

By plotting the variation with time of B[a]P and CO concentrations on the same graph (Figure 3), we can see 'by eye' that: (1) concentrations of the two pollutants vary throughout the day in a similar fashion; and (2) that this daily variation is consistent with variations in traffic volume – i.e. there are peaks during the morning and evening 'rush-hours', and a trough in the middle of the day.

So far, so good. From Figure 3 we can deduce that concentrations of B[a]P and CO are correlated and that their variation through the day is consistent with variations in traffic density, but we still do not know how strong this correlation is. This is

**Table 3**  Variation in concentrations of B[a]P and CO with time

| Time of day | B[a]P Concentration $(ng\ m^{-3})$ | CO concentration (ppmv) |
|---|---|---|
| 06:00–08:00 | 1.49 | 1.60 |
| 08:00–10:00 | 4.37 | 5.15 |
| 10:00–12:00 | 4.08 | 4.00 |
| 12:00–14:00 | 1.59 | 1.25 |
| 14:00–16:00 | 1.24 | 1.15 |
| 16:00–18:00 | 2.33 | 2.10 |
| 18:00–20:00 | 2.59 | 1.65 |
| Average | 2.53 | 2.41 |

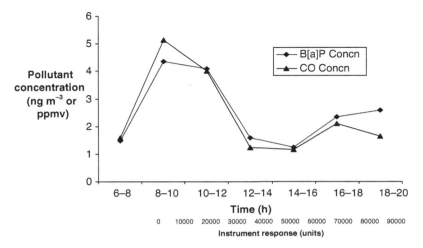

**Figure 3**   Variation of concentrations of B[a]P and CO during a given day.

important, as the stronger the correlation, the greater the contribution traffic makes to B[a]P concentrations. For example, if 40% of B[a]P is from sources other than traffic, then variations in emissions from these other sources (which will not affect CO concentrations), will reduce the strength of correlation between B[a]P and CO.

Consequently, the next step is to plot B[a]P concentrations as the dependent variable against the CO concentrations as the independent variable. By so doing, we obtain the graph shown in Figure 4.

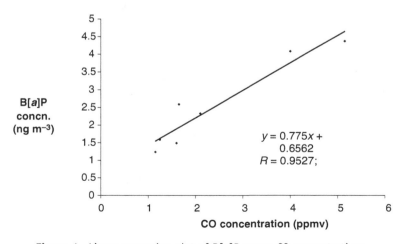

**Figure 4**   Linear regression plot of B[a]P versus CO concentrations.

The $R$ value is 0.9527 and $p < 0.001$. Thus we can say that the relationship between B[$a$]P and CO concentrations is linear at a statistical significance level in excess of 99.9% (i.e. there is a less than one in a thousand chance that the two variables are not correlated linearly) and that this relationship did not occur by chance.

The linear relationship between B[$a$]P and CO concentrations is expressed by the equation shown on the graph in Figure 4. We can subsequently use this equation to estimate the contribution of traffic to B[$a$]P concentrations by the following procedure.

From previous work, it is known that – in the absence of traffic – the concentration of CO is around 0.1 ppmv. Hence, by substituting $CO = 0.1$ into the linear regression equation relating B[$a$]P and CO, we can estimate the concentration of B[$a$]P in the absence of traffic, i.e. that due to all sources other than traffic, as

$$B[a]P = 0.775 \times 0.1 + 0.657 = 0.73 \, \text{ng m}^{-3} \text{(to three significant figures)}$$

Over the entire 06:00–20:00 hours sampling period, the average B[$a$]P concentration (i.e. that due to all sources including traffic) $= 2.53 \, \text{ng m}^{-3}$ (see Table 3). Thus, the average concentration due to traffic = the average concentration due to all sources – the concentration due to all sources other than traffic $= 1.80 \, \text{ng m}^{-3}$ (2.53 – 0.73), and the percentage of B[$a$]P due to traffic at this site on this day $= (1.80/2.53) \times 100 = 71\%$.

## 5.2 Multiple linear regression – to what extent is a given variable influenced by a range of other variables?

In the previous section, we examined how we could tell whether and to what extent values of one variable are dependent on values of another (independent) variable. As shown by the $R^2$ value in Figure 5.1.3, 32.57% of the variation in concentrations of PCB #18 is caused by changes in values of reciprocal temperature. The question that then arises is: what causes the remaining $100 - 32.57\%$, i.e. 67.43% of the variation in reciprocal temperature? This section addresses that sort of question. As mentioned in section 5.1, increases in temperature result in increasing concentrations of PCBs in air. Before we start to examine possible relationships between any dependent variable (in this case PCB concentration) and any number of possible independent variables, we need to think a little about the underpinning science of the situation we are considering and think what factors might influence logically the dependent variable. In the example here, we have already identified temperature as an influential independent variable. Other factors might be:

- wind direction in degrees relative to true north (if the wind at our measurement site passes over a large source of PCBs before it arrives, then we might expect higher concentrations on days when the wind blows from that direction);

- wind speed (if the wind speed is high, then we might anticipate PCB concentrations to be lower, as the increased air turbulence will dilute the PCB concentration);

- rainfall (higher rainfall could possibly 'wash' PCB contaminated particles out of the atmosphere).

As discussed in section 5.1, linear regression analysis identifies the best linear relationship between two variables of the form $y$ and $x$ and expresses it in the form of an equation of the type $y = mx + c$, where $m$ is the slope and $c$ the $y$-intercept. Multiple linear regression analysis (MLRA) is simply an extension of this principle enabling us to relate one dependent variable to a whole set of independent variables yielding a regression equation like that below:

$$y = m_1 a + m_2 b \cdots + m_n n + C$$

where $y$ is the dependent variable, $a$, $b$ and $n$ are the first, second, and $n$th independent variables; $m_1$, $m_2$, and $m_n$ are the first, second and $n$th slopes (or regression coefficients); and $C$ is the constant of regression – the MLRA equivalent of the $y$-intercept in linear regression analysis (see section 5.1).

In the current example, we would use MLRA to evaluate the linear dependence of PCB concentration on: (1) temperature; (2) wind direction (expressed as two separate variables – the cosine and the sine of the wind direction angle); (3) wind speed; and (4) rainfall.

When this is conducted you will obtain – as a first step – information about the relationship between PCB concentration and the above variables. This can be conducted in Excel, Minitab, and SPSS. Note that while the most recent versions of Excel can conduct MLRA – accessed via the Tools pull-down followed by Data Analysis and Regression (as mentioned before, note that you may have to install the Data Analysis Toolpak Add-In from your Excel or Office CD) – some earlier versions of Excel do not conduct MLRA.

It is by no means unusual at this stage to find that not all of the possible independent variables exert a significant influence on the dependent variable; indeed, making such discoveries is one of the aims of this initial step. What MLRA does is to allow you to identify such non-influential variables, and exclude them. The key information provided that allows you to identify whether an independent variable significantly influences the dependent variable is known as significant T in SPSS and P-value in Excel. It is effectively the MLRA equivalent of the $p$-value, and if it exceeds 0.1 for a given independent variable, then we can say with greater than $((1 - 0.1) \times 100 = 90\%)$ confidence that that independent variable does not influence the dependent variable.

**Table 5.2.1** Typical relevant multiple linear regression analysis (MLRA) output from SPSS

| Variable | B | β | Significant T |
|---|---|---|---|
| $1/T$ | −6323 | 0.8279 | 0.0002 |
| · cos *WD* | 0.23 | 0.5769 | 0.0044 |
| *C* | 8.6 | | 0.0003 |
| sin *WD* | | | 0.9996 |
| *RF* | | | 0.4763 |
| *WS* | | | 0.3337 |
| Overall significant F value = 0.0004. | | | |
| Adjusted $R^2$ value = 0.5345. | | | |

*T*, temperature; *WD*, wind direction; *RF*, rainfall; *WS*, wind speed.

Tables 5.2.1 and 5.2.2 summarise a typical MLRA output from both SPSS and later versions of Excel, for a MLRA of the relationship between PCB concentration and various independent variables.

Looking at the significant T or P-values, we see that those for: sine of wind direction (sin *WD*), wind speed (*WS*), and rainfall (*RF*) all exceed 0.1, while those for the constant *C*, reciprocal temperature ($1/T$), and cosine wind direction (cos *WD*) are all less than 0.1. Based on this, we can say that PCB concentration is **not** related linearly to the sine of wind direction, wind speed, or rainfall, but that it **is** related linearly to reciprocal temperature and the cosine of the wind direction. In other words, MLRA of our data reveals a linear relationship of the type:

$$\text{PCB concentration} = (m_1 \times 1/T) + (m_2 \times \cos WD) + C$$

The overall significance of this relationship is denoted by the Significant F value. In this case, it is less than 0.001, and we can say that the relationship is significant at the 99.9% confidence level. The values of the 'slopes' or regression coefficients

**Table 5.2.2** Typical relevant multiple linear regression analysis (MLRA) output from Excel

| Variable | Coefficients | P-value |
|---|---|---|
| $1/T$ | −6323 | 0.0002 |
| cos WD | 0.23 | 0.0044 |
| C | 8.6 | 0.0003 |
| sin WD | | 0.9996 |
| Rainfall | | 0.4763 |
| Wind speed | | 0.3337 |
| Overall significant F value = 0.0004 | | |
| Adjusted $R^2$ value = 0.5345 | | |

*T*, temperature; *WD*, wind direction; *RF*, rainfall; *WS*, wind speed.

($m_1$ and $m_2$), as well as the constant ($C$) are equal to the B values in SPSS or are reported as the coefficients in Excel, and allow us to derive the equation relating the dependent and independent variables. In this case, the relationship is:

$$\text{PCB concentration} = (-6323 \times 1/T) + (0.23 \times \cos WD) + 8.6$$

The last relevant piece of information supplied by a MLRA is the $\beta$ value. Unfortunately, this is only provided by more sophisticated statistical software such as Minitab or SPSS, but not Excel. The larger this value, the greater the influence on the dependent variable of a given independent variable relative to other independent variables. In other words, the relationship above does not tell us whether it is reciprocal temperature or cosine of the wind direction that is the most influential of the two factors in determining PCB concentration. However, the $\beta$ value for $1/T$ exceeds that for $\cos WD$, and we can therefore say that reciprocal temperature is the more important of the two independent variables. The other pertinent information yielded by both SPSS and Excel is the Adjusted $R^2$ value. Like the $R^2$ value in linear regression, this value expresses the percentage of the variability in the value of the dependent variable that is attributable to variations in the value of the independent variables in the MLRA equation. As shown in Tables 5.2.1 and 5.2.2, the value is 0.5345, and we can therefore say that 53.45% of the variation in PCB concentration is due to a combination of temperature and the cosine of the wind direction.

Having derived the relationship above between PCB concentration, reciprocal temperature, and $\cos WD$, we must interpret it in a way that enhances our scientific knowledge. It is all very well having a statistically significant linear relationship, but no good if we do not know what it means. We already know that increasing temperature increases PCB concentrations, but we now have additional information relating to $\cos WD$. The 'slope' or regression coefficient for $\cos WD$ in the above equation is a positive one, telling us that the PCB concentration increases when $\cos WD$ increases. Positive $\cos WD$ values are associated with wind directions of $270°$–$0°$–$90$, peaking at $0°$. We can interpret this as saying that PCB concentrations increase when the wind blows from a direction north of our sampling site. As our sampling site was south of Birmingham city centre, this implies that the city centre of Birmingham constitutes an important source of PCBs.

## 5.3  Non-linear regression

Although regression is a very useful tool for determining relationships between two variables, the basic model used assumes that the relationship is linear. However, in many cases that you come across the relationship may not fit this assumption and will be **non-linear** (many biological systems show non-linear relationships). In such instances we need to apply different tests in order to determine whether the relationship is significant or not. Remember that the typical linear regression is based upon a simple mathematical function that describes the slope of the line of best fit. When we

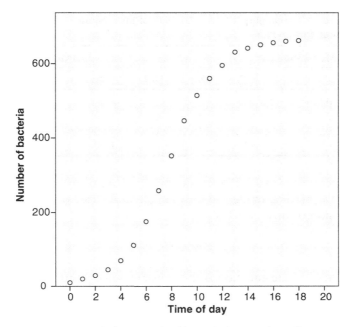

**Figure 5.3.1** Population growth of bacteria in growth media, pH 8.0.

move on to non-linear regression, then we are again applying mathematical models that are simply more complex.

Figure 5.3.1 shows a scatter plot of the population growth in bacteria in a growth medium. You can see clearly that although a relationship between the dependent and independent variables is apparent, it does not appear to be linear.

One of the first things that we can do when we come across such data is to log transform the dependent variable. Remember that with regression we want to establish whether we can predict what happens to the dependent variable when we change the independent variable, therefore it does not matter whether we use log transformed data. However, this does not always result in a linear relationship and in this case we may want to carry out non-linear regression.

We can use statistics to determine whether the curve does fit the linear relationship or whether it fits some other relationship (e.g. exponential, logistic). This can be achieved using something known as the 'curve estimation' and enables us to compare the relationship against a range of non-linear models. There are many different types of models available but it is outside the scope of this book to go into details and readers are referred to Miles and Shevlin (2000).

So if we take the curve from Figure 5.3.1 and apply a curve estimate for a number of models (linear, growth, exponential and logistical) we obtain the output from SPSS 15.0 (this can be found within the regression section of the program) shown in Table 5.3.1.

**Table 5.3.1**  Curve estimation output from SPSS 15.0 for linear, growth, exponential, and logistic models for bacterial growth data

| Equation | Model summary | | | | Parameter estimates | | |
|---|---|---|---|---|---|---|---|
| | $R^2$ | F | DF1 | DF2 | Sig. | Constant | $b_1$ |
| Linear | 0.935 | 243.210 | 1 | 17 | 0.000 | −36.648 | 45.521 |
| Growth | 0.832 | 84.391 | 1 | 17 | 0.000 | 3.371 | 0.222 |
| Exponential | 0.832 | 84.391 | 1 | 17 | 0.000 | 29.116 | 0.222 |
| Logistic | 0.968 | 517.745 | 1 | 17 | 0.000 | 0.057 | 0.659 |

The components of the table that we examine first are the parameter estimates and these relate to the original model. So for example in the case of the linear model the original expression of this model is:

$$y = b_0 + (b_1 \times t)$$

The $b_0$ component of the equation is the constant value given in Table 5.3.1. So in this case

$$y = -36.648 + 45.521$$

In the case of the logistic equation the model is:

$$y = 1/\{1/u + [b_0 \times (b_1 \times t)]\}$$

The $u$ component is the upper boundary value, which must be a number greater than the largest independent variable value. This is set before applying the curve estimation.

$$y = 1/\{1/700 + [0.057 \times (0.659)]\}$$

Because the $b_1$ value given exceeds 1 for the linear model this suggests a positive relationship between the two variables, so the greater the number of hours, the greater the number of bacteria. We can then examine the significance value (denoted by Sig. in Table 5.3.1), which in all cases is $< 0.05$. This tells us that the variation in the data explained by each model is not due to chance. The second component of the table we are interested in is the $R^2$ value that gives an indication of the strength of the relationship between the values that have been observed and those that were predicted by the model. All of the $R^2$ values are relatively large but the figure for the logistic model is largest (0.968) suggesting the strongest relationship, i.e. this is the model that best describes the observed data. The fits of the models can also be visually displayed as in Figure 5.3.2.

The exponential and growth models fall on exactly the same curve, but we can see clearly that the logistic model fits the measured data well, supporting the numerical results in Table 5.3.1. The interesting thing to note here is that the linear relationship

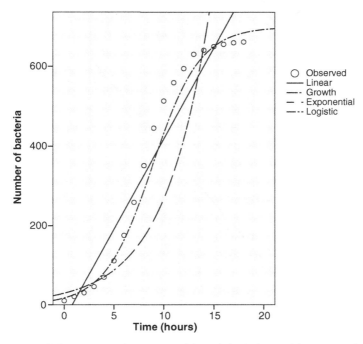

**Figure 5.3.2** Fit of linear, growth, exponential and logistic models to population growth of bacteria.

quite clearly does not fit the curve of the measured data. However, if we go back to the table the results of the curve estimation could (erroneously) be interpreted as suggesting that the linear relationship was a good fit (with an $R^2$ value of 0.935 and significance value $<0.05$). This highlights the importance of examining the curves visually, and also of selecting the appropriate model when testing non-linear regression. The positive results in the table for the linear model are probably due to the central section of the data, where there is a linear relationship. Indeed, if we carry out the analysis for curve estimation of this central section using a linear model then we obtain the following results (Table 5.3.2).

Now the $R^2$ value has increased to 0.997 and the significance level remains $<0.05$. Furthermore, if we examine the visual representation of the model (Figure 5.3.3) we can see that the linear model is a good fit for the central portion of the plot.

**Table 5.3.2** Curve estimation output from SPSS 15.0 for linear, growth, exponential, and logistic models for bacterial growth data (5–10 h)

| Equation | Model summary | | | | Parameter estimates | | |
|----------|------|------|-----|-----|------|----------|-------|
| | $R^2$ | F | DF1 | DF2 | Sig. | Constant | $b_1$ |
| Linear | 0.997 | 1171.591 | 1 | 4 | 0.000 | −315.052 | 83.271 |

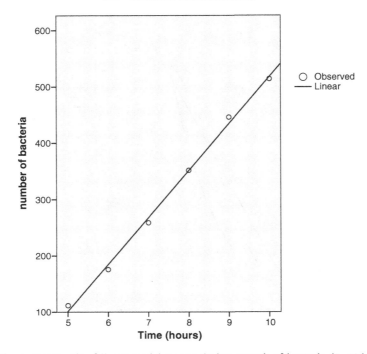

**Figure 5.3.3**   Fit of linear model to population growth of bacteria (5–10 h).

If the relationship within our data did not fit any of the basic non-linear models, we may require a more complex model that can also be tested in a similar way using non-linear regression. Students are referred to Miles and Shevlin (2000) for explanations of the types of models that may be used.

It should be remembered that there are assumptions for non-linear regression just as for any other type of statistical test. The dependent and independent variables should both be quantitative data. In addition any non-linear regression that you perform will only be valid if the function that is used (i.e. the model) describes accurately the relationship between the dependent and independent variables.

## 5.4   Pattern recognition

There may be cases in research where we are interested in establishing not whether there are relationships and differences between populations or variables, but instead whether there are patterns evident within datasets. For example, we may want to determine whether there are patterns to the distribution of vegetation types in response to environmental factors and if so which factors are the most important, or we may want to establish whether we can determine the relative impact on water quality of effluents from a number of different industrial sources. In both these cases we would be dealing with large sets of data involving many different variables, which necessitates the use of multivariate statistics.

If we have a dataset involving a large number of variables it is often difficult to make any sense out of it; what we need to do is reduce the complexity of the data so that we can isolate the most important factors that are influencing the patterns within the dataset. For example, we may want to determine whether vegetation type is influenced by particular environmental variables. In this case we may measure plant species growth and a number of other variables (e.g. pH, Ca content of soil, soil organic matter content, soil moisture, sunshine hours, air temperature, soil temperature, etc.). Once we have these data, we could go laboriously through every single variable and use univariate statistics to determine relationships between variables, however, a much more practical and robust approach is to use multivariate statistics.

One of the main statistical tools that we can use to achieve this is Principal Components Analysis (PCA). What it does is convert the original dataset which comprises measurements for each of the variable (e.g. pH) measured for each sample (often termed as a "case") and convert them to an equal number of composite variable. So far, so what, you may think. The neat bit is that usually the vast majority (typically 80% or more) of the variability in the dataset (in effect the useful information) is explained by two or three composite variables (known as principal components) rather than the far more numerous number of original variables. Hence, by plotting the values of these principal components for each sample, it is possible to check for patterns in the dataset in two or three dimensions, rather than 15 or 20 or so. Principal components analysis is not feasible without a computer and so must be carried out using a suitable statistics package such as Minitab or SPSS (Excel will not perform this test).

It is worth noting at this point that because the PCA does not actually 'test' anything but instead is a tool that identifies variation and patterns in data, it makes no assumptions about the distribution of the data. Thus as opposed to the majority of tests we have seen so far, we do not need to test the data for normality of distribution. The data, however, must be continuous and not ordinal (see section 2.1).

Let us take an example. A researcher wants to establish whether industrial activities are important sources of pollutants in house dust. One-hundred and fifty dust samples are taken from three locations (rural, urban and industrial) and analysed for ten different industrial pollutants. This generates a large dataset that is then subjected to a PCA using a suitable statistics package.

The computer will generate a number of different outputs for interpretation all related to the principal components. Each PC has something known as an eigenvalue, which is a measure of the variance within the population that the principal component accounts for. Table 5.4.1 gives the principal components and eigenvalues for our house dust example.

**Table 5.4.1** Eigenvalues of principal components analysis for house dust

|  | PC1 | PC2 | PC3 | PC4 | PC5 | PC6 | PC7 | PC8 | PC9 | PC10 |
|---|---|---|---|---|---|---|---|---|---|---|
| Eigenvalue | 4.2915 | 2.5627 | 0.7844 | 0.6670 | 0.5530 | 0.4351 | 0.3046 | 0.2249 | 0.1101 | 0.0666 |
| Proportion | 0.429 | 0.256 | 0.078 | 0.067 | 0.055 | 0.044 | 0.030 | 0.022 | 0.011 | 0.007 |
| Cumulative | 0.429 | 0.685 | 0.764 | 0.831 | 0.886 | 0.929 | 0.960 | 0.982 | 0.993 | 1.000 |

The values that we are interested in are the proportions and cumulative proportions of variance that each PC accounts for. So we can see that by using PC1 to describe house dust we would be describing 42.9% (0.429 × 100) of the variation between house types. We can describe 76.4% of the variation by using PC1, PC2 and PC3, which means that we only need three types of measurement rather than all 10 measurements. This particular result is not actually very good in that the variation is not concentrated hugely in the first few components but is distributed across a number of components. We can see this more clearly on a **scree plot** (Figure 5.4.1), which is a visual representation of Table 5.4.1. The components that fall on the steep part of the curve are the ones we are interested in, so we can see that it is the first three PCs that we are interested in, as they explain 76.4% of the variance. This means that if we look at these components only we will lose 23.6% of the information from the dataset. If we include PC4 as well then we will only miss 16.9% of information and therefore it may be worth including this component in further analysis of data.

So now we need to examine the **eigenvectors**, which tell us how close each principal component is to the original variables. The eigenvectors for our house dust data are given in Table 5.4.2. Remember that we have just concluded that it is only PC 1, 2, 3 and 4 that we are really interested in as they explain most of the variance. Eigenvector values close to 0 indicate that there is little relationship between the scores on the PCA and the original variables. So we can see from Table 5.4.2 that scores on PC1 are very poorly related to Ni and Cd. The two variables that seem to be most closely linked to PC1 are those with the highest eigenvectors: Al (0.406) and S (0.457). We also need to compare these values to the equivalents in

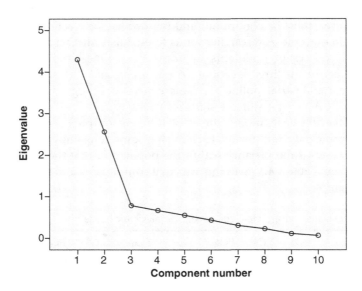

**Figure 5.4.1** Scree plot for the principal components for house dust.

**Table 5.4.2** Eigenvectors for all components

|    | PC1 | PC2 | PC3 | PC4 | PC5 | PC6 | PC7 | PC8 | PC9 | PC10 |
|----|-----|-----|-----|-----|-----|-----|-----|-----|-----|------|
| Al | 0.406 | −0.162 | −0.043 | −0.029 | 0.232 | 0.405 | −0.540 | 0.150 | −0.508 | 0.126 |
| Cd | 0.039 | 0.587 | 0.052 | 0.067 | 0.129 | −0.105 | 0.279 | 0.381 | −0.476 | −0.412 |
| Co | 0.329 | −0.109 | 0.335 | 0.342 | −0.767 | 0.054 | 0.041 | 0.240 | −0.030 | 0.005 |
| Cu | 0.305 | −0.059 | −0.476 | −0.690 | 0.367 | −0.174 | 0.143 | 0.032 | −0.115 | −0.037 |
| Fe | 0.364 | 0.189 | 0.309 | −0.003 | 0.066 | −0.660 | −0.295 | −0.445 | −0.101 | 0.034 |
| Ni | −0.087 | 0.500 | 0.269 | −0.301 | −0.238 | 0.538 | −0.022 | −0.480 | −0.022 | 0.043 |
| Pb | 0.160 | 0.344 | −0.690 | 0.526 | −0.130 | 0.055 | −0.097 | −0.254 | 0.075 | 0.060 |
| S  | 0.457 | −0.020 | 0.049 | −0.049 | 0.186 | 0.201 | −0.075 | 0.008 | 0.537 | −0.645 |
| V  | −0.356 | 0.289 | −0.051 | −0.153 | −0.221 | −0.152 | −0.690 | 0.386 | 0.244 | −0.076 |
| Zn | 0.361 | 0.355 | 0.088 | −0.094 | 0.214 | 0.021 | 0.168 | 0.362 | 0.370 | 0.620 |

PC2 which are −0.162 and −0.020 respectively, suggesting that S is the most important as it is also unrelated to PC2 (low eigenvector). Both eigenvector values for Al and S in PC1 are strongly positive telling us that house dust samples which have high scores on PC1 will have high concentrations of Al and S. For PC2 the two variables most closely linked are Cd and Ni, for PC3 it is Pb and Cu (both negative values in this case meaning that house dust with high scores for PC3 will have low concentrations of Pb and Cu), and for PC4 it is Cu and Pb again but with a positive value for Pb.

From the computer package you will also be able to obtain values which reflect how each house type relates to the different PCs. We originally took 150 dust samples comprising 50 from houses in each of three areas (one a rural location, one an urban location and one in the vicinity of an industrial processing plant). Table 5.4.3 gives example scores for each of the three areas.

As before we are looking for the largest values (whether they be negative or positive). So for PC1 we can see that the largest values can be found in samples from rural locations, but note that these values are negative. This means that house dust from rural locations is characterised by low concentrations of Al and S. It is also evident that there are high scores for the industrial sites but in this case they are positive and therefore characterised by high concentrations of Al and S. We can do the same for the other PCs, so for PC2 the highest (positive) values are found in urban sites and thus driven by high concentrations of Cd and Ni. We can actually test whether there is a significant difference between values for different PCs for the different sample categories using statistical tests such as t-tests or ANOVA if required.

So now from carrying out PCA on this dataset we can identify typical characteristics of house dust from areas that receive different types and amounts of contaminants. We could then take samples from local pollution sources (such as an aluminium smelter) and use PCA to establish whether there are similarities between the contaminant patterns in the source and the house dust. If so, then this would provide information about the sources of the indoor dust.

**Table 5.4.3** Principal component scores for selected house dust samples

| Area | Sample | PC1 | PC2 | PC3 | PC4 |
|---|---|---|---|---|---|
| Urban | 1 | 0.032 | 0.953 | −0.599 | −1.327 |
| | 2 | 1.882 | 1.983 | 0.066 | 0.0959 |
| | 3 | 1.016 | 1.984 | 1.952 | −0.254 |
| | 4 | 1.699 | 2.161 | 0.581 | 0.958 |
| | 5 | 0.141 | 0.956 | −0.10 | −2.776 |
| Rural | 1 | −2.575 | −0.822 | 0.512 | 0.392 |
| | 2 | −1.971 | −1.035 | 0.204 | 0.922 |
| | 3 | −2.980 | −0.780 | 0.614 | 0.944 |
| | 4 | −3.202 | −1.367 | −0.476 | −0.354 |
| | 5 | −2.393 | −1.243 | 1.161 | 1.297 |
| Industrial | 1 | 2.452 | 0.494 | 1.306 | −0.953 |
| | 2 | 2.418 | −0.793 | −1.007 | −1.055 |
| | 3 | 0.507 | −1.022 | −0.355 | −0.485 |
| | 4 | 1.980 | −0.503 | 0.0516 | 0.311 |
| | 5 | 2.118 | −0.568 | 0.463 | −0.532 |

There is still some further information that we can obtain from PCA. We can use it to determine whether there are any distinct groupings of individuals within the dataset. We can do this manually by plotting each variable against the other in turn. However, given that we have 10 variables, this would generate many graphs, become extremely complex to interpret, and only use the information from two variables at a time. An alternative is to use our principal component scores (sometimes referred to as factor scores) from the PCA. This means that we are using information from all the variables (remember each PC is a composite of all of the original variables). We have already established that we are only interested in the first four principal components and so we will only generate scatterplots for these (Figure 5.4.2).

You can see quite clearly in the first scatterplot that the values appear to fall into three distinct groups. Those samples that occur towards the bottom left of the graph have negative values for both PC1 and 2 (they have low concentrations of Al, S, Cd and Ni) and therefore correspond to the rural sites. Those in the bottom right of the graph have negative values for PC2 and positive values for PC1 (low concentrations of Cd and Ni, high concentrations of S and Al) corresponding to the industrial sites. The remaining grouping matches the urban site characteristics. You may be able to see groupings in some of the other plots that you can interpret in the same way. This type of analysis can aid us in identifying groupings within large datasets that can be used for further analysis, or to provide information for use in future sampling campaigns.

We have so far examined ways of using multivariate statistics to determine which variables are most important in any population using PCA. Alternatively, if we have measured a number of variables in a population we may be interested in identifying whether there are 'clusters' of individuals within the population, i.e. do certain individuals have similar characteristics in terms of the measured variables. So, for example, we may be interested in determining whether certain plant species require

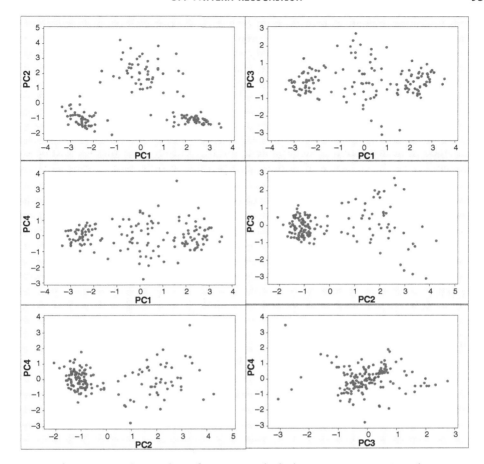

**Figure 5.4.2** Scatterplots of scores on principal components 1, 2, 3, and 4.

different environmental conditions. We can collect data on the occurrence of different plant species under field conditions together with a range of environmental data such as pH, moisture content, organic matter content, altitude, slope, Ca content, cation exchange capacity, etc., in order to determine whether certain species require particular conditions. In order to do this we require a statistical method known as **cluster analysis**.

There are a range of different approaches to cluster analysis which differ in the way similarities in data are defined. Single linkage is the most widely used approach, and we will thus focus our discussion primarily on this, while referring briefly to the other possible approaches. Readers wishing to know about these are referred to Everitt *et al.* (2001).

- *Single linkage*. This analysis identifies the minimum distance between a variable within one cluster and a variable in another cluster and forms a link between them.

This is an appropriate clustering method to use if your clusters are well defined. The problem with this approach is that when variables are close together and clusters poorly separated, using single linkage may cause something known as **chaining**. This results in individuals being linked to the nearest cluster in a chain. However, this is the best approach for initially identifying clusters and can demonstrate clearly the absence of clusters.

- *Complete linkage.* The links are formed between a variable in one cluster that is at the maximum distance from a variable in another cluster. This method can be very susceptible to the presence of outliers and produce groups where none exist.

- *Average linkage.* Linkage between pairs of objects depends upon the average distance and is essentially intermediate between complete and single linkage. This gives a central measure of location.

- *Centroid linkage.* This uses the centroid of each cluster (the mid-point) and forms a link between them. This often produces similar results to the average linkage method.

We can illustrate the use of cluster analysis thus. A researcher has collected soil chemistry data from a number of industrial sites and wants to establish whether sites can be classified (or clustered) according to the pollutants found within the soils. Cluster analysis based on single linkage is used initially in order to establish whether there are any clusters. A cluster analysis will produce a figure known as a dendrogram as shown in Figure 5.4.3.

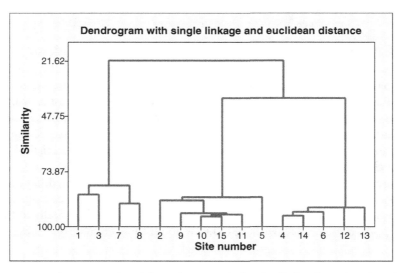

**Figure 5.4.3** Dendrogram produced from cluster analysis of soil chemistry data from industrial sites using single linkage with the variables standardised.

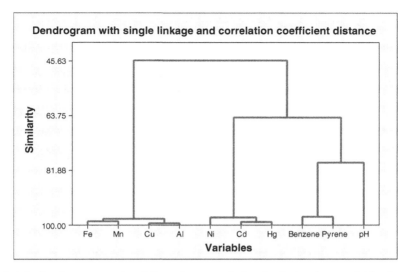

**Figure 5.4.4** Dendrogram produced from cluster analysis of soil chemistry variables from industrial sites using single linkage.

The figure quantifies the similarity between the different sites according to the variables (similarity is on the $y$ axis). We can see from the figure that there appears to be three distinct groups or clusters: sites 1, 3, 7 and 8; sites 2, 9, 10, 15, 11 and 5; and sites 4, 14, 6, 12 and 13. These groups are no more than around 40% similar, which we can see by looking at the level where they split off. We can then look back at our dataset to establish which of the variables may be influencing the clusters. Alternatively we can carry out another analysis in which we can cluster the data according to the variables, rather than the single observations. In this case we obtain the dendrogram in Figure 5.4.4.

We can interpret the dendrogram in the same way, such that it shows there to be three clusters: Fe, Mn, Cu, and Al; Ni, Cd, and Hg; benzene, pyrene, and pH. We may infer that the clustering shown in the sites (Figure 5.4.3) may be due to the differences identified in the magnitudes of the chemical variables identified in Figure 5.4.4. Further analysis of the data using other statistical methods such as ANOVA can then be used to explore this.

## Recommended reading

Bryman, A. and Cramer, D. 1996. *Quantitative Data Analysis with Minitab. A Guide for Social Scientists*, Routledge, London. ISBN: 9780415123242.

Everitt, B.S., Landau, S. and Leese, M. 2001. *Cluster Analysis*, 4th edn, Hodder Arnold, London.

Grimm, L.G. and Yarnold, P. 1995. *Reading and Understanding Multivariate Statistics*, American Psychological Association, Washington DC.

Johnson, R. and Wichern, D. 1992. *Applied Multivariate Statistical Methods*, 3rd edn, Prentice Hall, Harlow.

Keller, G. 2001. *Applied Statistics with Microsoft*® Brooks Cole, Pacific Grove, CA, USA. ISBN: 0534371124.

Kinnear, P. R. and Gray, C. D. 2007. *SPSS 15 Made Simple*, Psychology Press. ISBN-10: 1841696862.

Lindley, D.V. and Scott, W.F. 1995. *New Cambridge Statistical Tables*, Cambridge University Press, Cambridge.

Manly, B.F.J. 2004. *Multivariate Statistical Methods: a Primer*, 3rd edn, Chapman & Hall, London.

Miles, J. and Shevlin, M. 2000. *Applying Regression and Correlation: A Guide for Students and Researchers*, Sage Publishers Ltd, London.

Rencher, A.C. 1995. *Methods of Multivariate Analysis*, John Wiley & Sons, Chichester.

# 6

# Making sense of past, present and future systems – mathematical modelling

## 6.1  What is a model?

In the morning when you rise, you use a 'mental' model to get out of bed and to navigate through the day. Your mental model enables you to decide what to eat for breakfast (I am in a hurry therefore I had better eat some toast), how to get from point A to B (I need to get to the university quickly but I am adverse to taking too many risks so I will not take my unicycle today), and how to achieve your goals (I need to stay awake to write an assignment so I will drink a large coffee but I will not drink a jumbo large since then I will lose time as I will get too jittery). The models we all construct in our minds are our explanations of how the 'real' world works – they are our unique representation of external reality. Our mental models contain numerous implicit assumptions (e.g. personal success is defined by contentment), and facts about yourself (e.g. my aversion or thrill for risks directs whether I read a book or go bungy jumping) and your environment (e.g. the world is an exciting place to explore or the world is a dangerous place that requires constant vigilance).

In environmental science we use models constantly to organise and explain phenomena (e.g. I observe $X$ and $Y$, which implies $Z$; or, if I modify the environment by doing $X$ then $Y$ happens, which is consistent with my hypothesis $Z$). **A model is simply a theoretical construct that represents something**. It is an analogy. A model is comprised of a set of logical relationships set within an idealised framework or system. The relationships can be **qualitative** (e.g. our mental models, conceptual models) or **quantitative** as in mathematical models. The idealised and logical framework is the model builder's simplification of reality, where the simplifications are based on a series of assumptions that are known to be false or incomplete. These simplifying assumptions are necessary in order to allow

*Student Projects in Environmental Science*   Edited by Stuart Harrad and Lesley Batty
© 2008 John Wiley & Sons Ltd

us to develop, use, and understand the model. In short, **a model is a structured and disciplined framework that simplifies reality and provides an 'argumentative framework'.**

This brings us to the purpose of models. Models have diverse uses, as discussed below. They allow us to test hypotheses, to improve our understanding of systems, and to predict future scenarios, to mention a few (e.g. to predict that an apple will fall from a tree according to the law of gravity). The very act of developing a model forces you to examine logically the system or process you are investigating and as such, to better understand the system and its complexities. Developing a model disciplines you to organise your information and to identify model components, dependent and independent variables, causative agents, and relationships among model components. Indeed, model development itself may increase your understanding of the system or allow you to identify faulty logic or suppositions.

Models are used extensively in environmental science. As scientific tools, they are popular because models can reduce the complexity seen in ecosystems, thereby allowing you to probe system behaviour and relationships within the system. They can be used to conduct 'experiments' that would otherwise be impossible to do. Models can be used to simulate past and/or future events where the future can be manipulated by assuming different scenarios. Models also have widespread use in environmental decision-making and management. We can use models to explore 'what-if?' scenarios corresponding to different management options. Model output is used frequently to communicate preferred options to managers and/or politicians who must make decisions based on limited knowledge.

Environmental models must work from a firm foundation of knowledge and data that describe the system. As such, modellers often work in teams with other environmental scientists who conduct field work, and collect and analyse data. It is often tempting to overstate the power of models, as can occur when modellers are divorced from those with an intimate understanding of the complexities and knowledge gaps of the system and relevant data. Examples of models used in environmental science include those for air pollution chemistry, relating air pollution to adverse health effects, hydrology, surface and ground water quality, population dynamics (e.g. fisheries models), nutrient cycling, ecosystem dynamics, climate change on global, regional and local scales, and system adaptation, to name a few. Indeed, models may be the only means of understanding complex systems for which measurement and observation are insufficient.

In this chapter we discuss the assortment of models that can be developed or are available, the functions that models play, and the components of a model. From there, we examine how models are developed and evaluated. We end by illustrating the use of models with a detailed case study. The discussion can guide you to develop your own model or to better understand an 'off-the-shelf' model or one that you have downloaded. If it is the latter, you need to know how the model works in order to use it appropriately. Finally, before we embark on this discussion, remember that models are not only informative but can be fun to use in the same way that computer games, which are types of models, are so engaging.

## 6.2  Functions of models

As mentioned above, models have numerous uses in scientific and management arenas. In the scientific arena, Hartmann (1996) summarises five functions of model simulations. We believe that these functions apply broadly to models used in environmental science. The five functions that Hartmann lists are:

- as a technique for investigating detailed dynamics of a system;

- as a heuristic tool for developing hypotheses, models, and theories;

- as a substitute for performing experiments;

- to support experimentalists;

- as a pedagogic tool that can be used to gain understanding of a process or system.

These functions are not mutually exclusive and thus often overlap. We discuss each in turn below.

Models can be an invaluable tool and resource for **investigating properties, processes and relationships in real world situations**. We can use models to explore components, or the entirety, of a system. In some cases the system may defy measurement because of its intractable scope (spatial, temporal, or in terms of the number and/or complexity of interacting components). An example of this are global circulation models (GCMs) that simulate global-scale climate over long temporal scales. These models are used extensively for scientific and policy purposes, such as negotiating for reductions in national carbon emissions. These models are constructed from numerous sub-models that reproduce integrated components of the global energy balance, such as two- and three-dimensional ocean circulation models, land–air energy balance with and without vegetation as a dynamic sub-component, and hydrological cycling. A GCM can be used to run 'what-if' scenarios ranging from 'business as usual' to drastic reductions in carbon emissions. Modellers can investigate sub-components of the model to evaluate the magnitude and time-frame of the response of the component to climate perturbations (e.g. the effect of oceanic circulation to the melting of Arctic glaciers). At the other extreme, a model can be constructed to simulate chemical movement through an organism (a pharmacokinetic model). If theoretically robust, the model will enable toxicologists and pharmacologists to estimate the time, course, and extent of the distribution of a chemical through a single organism without actually exposing the organism to the chemical.

The second function of models is as a **heuristic tool** to improve our understanding of the system. We may use the model to formulate a hypothesis, better understand a causative relationship(s), or to understand better the role(s) of various factors on system behaviour. The lack of correspondence between measured responses and modelled estimates may provide a clue that we do not understand fully the system behaviour or that the mathematical expression used is incorrect. If the modelling analysis shows that we do not understand the system, we can use the modelling results to direct research priorities.

Models can be used as an **experimental tool**. It may not be possible to conduct experiments at an ecosystem scale or to perturb a system in a particular way. In these cases, we can use a model of the system as our experimental tool. The experiments may consist of changing components of the system, values of variables, or any part of the model. We can evaluate system responses over time and/or space. Running carbon reduction scenarios on a GCM can be seen as global-scale experiments in climate change.

In addition to being an **experimental tool**, models can be used to support experimentalists. Models can be used to interpret and analyse observed data or to identify components for further investigation.

The final use of models discussed by Hartmann (1996) is as a **pedagogical tool**. Computer models are analogous to a computer game where you can 'play' with the system and observe its responses to your choices. You can develop an understanding of system behaviour that can lead to you predicting outcomes to newly explored scenarios.

The other major arena in which models are used is **decision-making** involving environmental managers, regulators, or politicians. Models have played an integral role in managing environmental systems, such as lake eutrophication (Dillon and Rigler, 1974; Vollenweider, 1975) and fisheries management (Hewett and Johnson, 1992). Often the systems to be managed are too large or complex to understand experimentally. Moreover, managers need tools to prioritise actions or choose among options and to justify decisions. It is here that models invaluably portray scenarios and allow informed or at least explicit choices to be made. The choices may not always be right, but models increase transparency and provide for an explicit acknowledgement of assumptions and trade-offs that accompany the decision.

In decision-making, models can have the following functions:

- displaying results of 'what if?' scenarios such as different management options;

- identifying data and/or knowledge gaps, thus prioritizing further data gathering efforts or study;

- identifying dominant processes or factors that require attention or monitoring;

- to probe uncertainties in knowledge of system behaviour;

- to fill in essential data gaps such as values of difficult-to-measure variables based on a sound theoretical understanding of the system.

## 6.3 Which type of model should I use?

Since there are many types of models, which model should you choose for your purposes? Figure 6.3.1 summarises a 'taxonomy' of models. The first choice is between a qualitative or conceptual model versus a quantitative or mathematical model. Actually, this may not be a choice since a qualitative model is a precursor to all mathematical models.

We started the chapter by discussing **mental models**, which each of us constructs as a cognitive representation that extends from an idea to our entire world. These are implicit **qualitative models** that allow us to operate on a daily basis. Implicit in our mental models are our ethical principles, political outlook, personal preferences, etc.

Moving into the scientific realm, we start all modelling efforts with a **conceptual or qualitative model**, which allows us to define the system that we are investigating – the components of the system and their interactions. The system is circumscribed by a **system boundary** that is defined as an arbitrary limit that distinguishes between the components of interest versus those excluded from our considerations.

A very simple conceptual model is 'A causes B' (e.g. phosphorus added to a lake increases primary productivity; or, exposure of an organism to a toxic dose of chemical causes death). Underlying these simple causal statements are complex sets of interactions amongst many components. In the case of phosphorus added to a lake, we can consider chemical interactions related to the chemical species of phosphorus that is added leading to its bioavailability, the availability of other essential nutrients such as nitrogen, temperature that affects algal growth rates, etc. In the second case, we can consider the transport of the toxin within the organism, competing reactions of the toxin with other biochemical constituents, repair mechanisms that can reduce the effect of the toxin, etc. The second set of descriptions can be translated into more complex models. Figure 6.3.2 illustrates a more complex conceptual model of a food web or 'who eats whom'. This conceptual model is a necessary precursor to developing a quantitative model of energy, nutrient, or contaminant transfer through this food web.

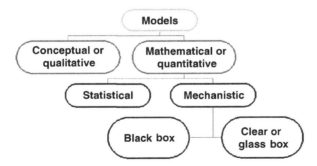

**Figure 6.3.1** Taxonomy of models.

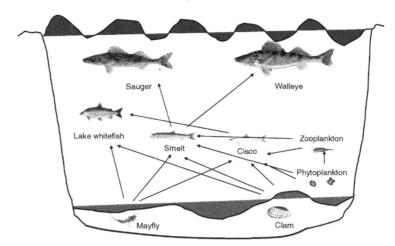

**Figure 6.3.2** Example of a conceptual model of a food web or 'who eats whom'. (Adapted from Gewurtz *et al.*, 2007. Dynamics of PCBs in the food web of Lake Winnipeg. *J. Great Lakes Res.*, **32**, 712–727.)

Sometimes systems are so complex or poorly known that we do our best just to draw a conceptual model. This may be the case with social systems where the system components may be countries, social groups, or activities. We illustrate this case using urban agriculture as an example. Figure 6.3.3 summarises the components and factors of urban agriculture that ought to be considered in order to develop policies to maximise benefits to urban farmers and their families. We have an individual situated within her/his family and community. Urban agriculture can confer many positive outcomes that prevail only when the negative factors are mitigated by risk management interventions. The conceptual model allows us to articulate the components of the system (individual, family, community) and both positive and negative factors impinging on those components.

We can use this very general model to guide our understanding of a particular situation – that of children's health and lead. In the context of urban agriculture, we are concerned with lead as an additive of petrol and its emission nearby roadways to lands that may be used for urban agriculture. Lead is a potent neurotoxin that affects learning and behaviour. Let us consider the case of urban agriculture in poor communities where food from urban agriculture may mean the difference between a full and empty stomach. As seen in Figure 6.3.4, the primary determinants of children's health are their parents' education and socio-economic status. If both are high, then the children may have a high socio-emotional status which leads to well-being through improved housing quality and location, improved food security and availability, and hence nutritional status of the children. Children with a higher socio-economic status are less likely to have neurological and behavioural problems than children with the same blood lead level but with a lower socio-economic status (Bellinger *et al.*, 1988). In contrast, children with elevated blood lead levels and low socio-economic status will tend to have lower educational attainment that will affect

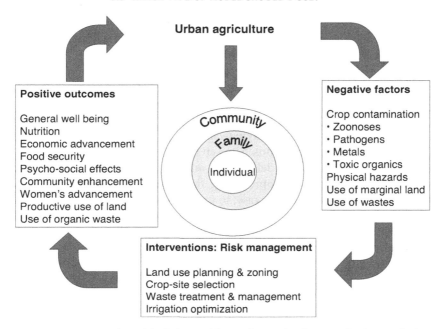

**Figure 6.3.3** A conceptual model of the positive and negative factors of urban agriculture that influence the health and well-being of an individual.

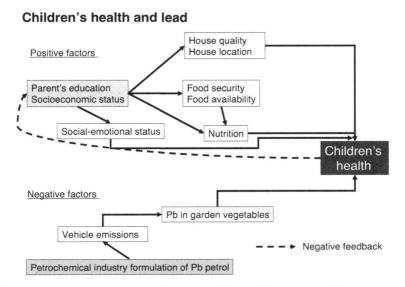

**Figure 6.3.4** A conceptual model of the relationships of children's health to socio-economic factors and exposure to lead from leaded petrol via consumption of vegetables grown in soils adjacent to heavily trafficked roadways.

adversely their children and so the cycle continues. From this model we can identify management actions aimed at improving children's health of eliminating lead from petrol and increasing parental socio-economic status.

This conceptual model is built from a review of numerous studies (e.g. Hertzman and Weins, 1996; Bernard and McGeehin, 2003). As such, the processes and cause-effect relationships have been well documented. However, it would be an enormous task to translate the conceptual cause–effect relationships into a quantitative model. Part of the difficulty comes in quantifying variables such as social-emotional status of the children, and the quantitative link between that status and the parent's education. Nonetheless, the conceptual model can be an important tool for understanding possible interrelationships and appreciating the breadth and intricacies of system components that can lead us, in this case, to productive interventions to minimise the adverse effects arising from exposure to lead in the context of urban agriculture.

The alternative to the conceptual or qualitative model is a **quantitative** or **mathematical** model (Figure 6.3.1). Chapter 5 discusses **statistical models**, which are commonly used quantitative models. Statistical models quantify the relationship (s) between **independent** (predictor, explanatory or controlled variables) and **dependent** (response) variables. Causality is usually assumed, but there is nothing in the model structure that demands it. Hence, one can run the risk of inferring a spurious relationship between the independent and dependent variables (see section 4.1). One could regress the loss of global biodiversity (dependent variable) against the rise in the number of computers (independent variable) over the past half century. Both variables are linked through time, but are not necessarily causally related (although one could develop a conceptual model relating these variables in the context of sustainable development!). A very common use of statistical models is in toxicology where we use statistically defined dose–response relationships between exposure (e.g. intake of chemical $X$) and physiological response (e.g. neurotoxicity, death). Dose–response relationships invaluably underpin applied toxicological and ecotoxicological tools such as hazard and risk assessment.

**Mechanistic models**, in contrast to statistical models, are built on causal relationships defined through mathematical expressions. The models synthesise the science of what we know about a system (Jorgensen and Bendoricchio, 2001) and the art and science of translating that knowledge into a simplified and clear mathematical framework. In section 6.4 we discuss the trade-offs between developing simple versus complex models.

We can develop several types of mechanistic models. Fundamental to mechanistic models is the **Lamonosov–Lavoisier Law of Conservation of Mass**. This law demands that mass remains constant within a **closed system**, where the latter is defined as a system with a boundary across which only heat and work (energy transferred by force), but not mass, can be transferred. Environmental modellers, however, deal with **open systems** in which mass does transfer across the system boundary as inputs and outputs. We still use the Law of Conservation of Mass to track the inputs and outputs across the boundary, i.e. all mass must be accounted for in the system.

Examples of mass balances include: biogeochemical models, where mass refers to an element such as carbon, nitrogen, or other nutrients; hydrological and hydrogeological models that track surface and groundwater respectively; and environmental fate and transport models and pharmacokinetic models that track chemicals in environmental and organismal systems, respectively. The energy conservation principle underpins bioenergetic models in which energy rather than mass is balanced. Climate models also track energy. We discuss the development of a mass balance model in section 6.4.

The term **material balance** is also used to track complex materials such as a plastic, mineral alloy, etc., using, for example, a material flow analysis (Daniels and Moore, 2002). Mass and material balances are also used in life cycle assessment, where we track inputs and outputs of chemicals, materials, and energy across the boundaries of each life-cycle stage (e.g. raw material extraction, manufacturing) that are within a larger boundary of a product or processes. We then use a combination of mechanistic and statistical models to relate these inputs and outputs to a variety of environmental impacts in order to estimate the environmental impact of the product or process (e.g. Fava *et al.*, 1993). In these types of models, the terms stocks and flows are commonly used, where the **stock** is mass of chemical or material in the system or a compartment of the system and the **flows** are the movement (e.g. inputs and outputs) of the chemical or material.

**Black box models** quantify inputs and outputs across a system boundary (Figure 6.3.5). We are familiar with such input–output models as we budget our money, hoping that our monetary inputs exceed our outputs. The processes relating inputs and outputs are absent in a black box model, i.e. we do not consider explicitly how or why an input results in an output, where **process** is defined as a phenomenon causing a change in one or many system properties. In other words, we do not have independent or explanatory variables in the model. An example of a black box model is a chemical budget for a lake (Figure 6.3.5A). We choose a lake because we can easily define and measure an input through the inflow and an output through the outflow. For this black box model we must measure both input and output. In this case, we assume inputs and outputs to and from the atmosphere and groundwater are negligible. We neglect the processes affecting the chemical within the lake such as sedimentation, resuspension, diffusion, etc. The model tells us the status of the system – whether the lake is a net sink or source of chemical. We cannot use the model to understand the factors affecting its status as a sink or source, nor can we use the model to make predictions.

The next level of complexity is a **clear** or **glass box** model. In contrast to the black box model, we now consider processes within the system boundary. Returning to the lake, a clear or glass box model considers processes of sediment–water exchange, atmospheric exchange, and chemical transformation as factors causing the output (Figure 6.3.5B). We can get into great mechanistic detail using multiple explanatory variables for each process. This category of model requires far more understanding of the system, model variables, and measurements to support model parameterisation. The advantages of this model are that it can be used to diagnose factors related to the

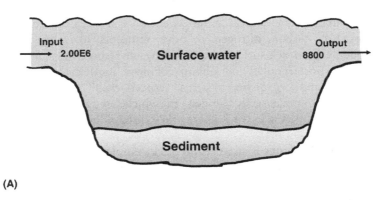

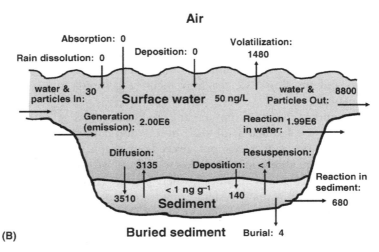

**Figure 6.3.5** An example of quantitative models of a chemical entering and leaving a lake. The chemical may be a nutrient or contaminant. (A) is an example of a 'black box', budget or input–output model; (B) is an example of a 'clear or glass box', or mechanistic model. These examples come from a steady-state model of the chemical geosim (produced by plankton that confers bad taste and odour to water) in Lake Ontario (Bhavsar, Schryer and Diamond, unpublished data). All rates are in units of g day$^{-1}$.

lake's status as sink or source and can be used as a predictive tool. Thus, we can use the model to run 'what-if' scenarios as we vary inputs and/or the values of model variables. As such, we can use this tool to assist with lake-wide planning and management.

Quantitative or mechanistic models can be run as either deterministic or probabilistic simulations. **Deterministic models** behave predictably – if you specify input *A* you will always get output *B*, i.e. the model goes through the same sequence of events every time it is run with the same inputs. Practically, this means that you construct your model with one set of values for your variables and the model produces reliably one set of outputs or model estimates for one set of inputs. A deterministic model is very useful for allowing us to understand how the system works, to 'experiment' with the system using the model (e.g. I vary $A$ which results in $Q$, I vary $B$ which results in $R$), and the critical dependent variables controlling the model's output or the **sensitivity of model** results to variations in the values of particular variables.

However, natural phenomena are seldom deterministic, although we often wish they would be (e.g. I will work to produce an outstanding thesis which will result in me getting an excellent job). Rather, there are elements of **random or stochastic** behaviour (e.g. I will work hard to produce an outstanding thesis which will give me a chance at getting an excellent job). Sometimes we can assign a **probability** to this randomness (e.g. if I produce an outstanding thesis then I have a higher probability of getting an outstanding job than if my thesis is lousy). This gives rise to **stochastic models** in which an input of $A$ will result in a range of outputs of the dependent variable $B$. The stochasticity is generated by **stochastic or random independent variables** within the model that can have a range of values. The stochastic model includes a specified a range of values that the random variables may have, e.g. from a low of $Q$ to a high of $Z$. The model, using a random number generator in a computer algorithm, will then select random values from within this range to give a range of outputs (the opposite of a deterministic model). If you know or can guess the statistical distribution of the range of values of the random variable, then you can generate a statistically defined probability distribution and your **probabilistic model** generates a probabilistic range of outputs. Stochastic or probabilistic models are used for **uncertainty analyses** whereby you define a range of values of independent variable(s) based on your incomplete knowledge of the exact value or exact range of such values. Stochastic and probabilistic models are also useful for expressing the range of model outputs based on the uncertainty and/or naturally occurring variability in independent variables. It provides decision-makers with a likely range of outcomes and a sense of the implications of uncertainty in model variables. The disadvantage of these models is their mathematical and computational complexity. As well, the results can be more difficult and time-consuming to interpret than results from a deterministic model.

We have discussed the basic types of models that you can build or that are available. There are additional features of mechanistic models that we need to know. As discussed in section 6.4 below, a mass balance model is set as an individual or series of differential equations where the mass is differentiated with respect to time

$$dM/dt = \text{inputs} - \text{outputs}$$

where $M$ is mass and $t$ is time. We solve the differential to obtain model output as a function of time. The solution provides us with a **time dependent, dynamic or unsteady-state model** (these terms are used synonymously). If the equation is relatively simple, then we can solve the equation mathematically to obtain an **explicit analytical**

**solution** where infinitesimal time-steps are implied. However, often the equation(s) are difficult to solve, in which case we use **numerical approximations**. The simplest of these is a **finite difference** approach whereby the computer algorithm steps through time according to defined and discrete time-steps (the time-steps are usually very small so that the model output is mathematically stable). We also need boundary conditions to solve the differential equations, such as the initial value at the start of the simulation, e.g. at $t = 0$. Other boundary conditions may also be necessary.

An alternative solution to solving the differential equation explicitly is to assume no change in the mass within the system with time or that the system is at **steady state**, i.e. $dM/dt = 0$. If we make this assumption, then the mass balance solution becomes the more easily solved equation:

$$\text{inputs} = \text{outputs}$$

Specifically, we assume that there is no change in mass with time as we force the inputs to equal the outputs. Examples of steady-state systems are a constantly flowing river or water fall. Note that the system is not necessarily static, but rather can be dynamic with constant rate(s) of movement, transport, reaction, growth, etc., i.e. the rates do not vary over time. While we are at this point, let us distinguish between a system at steady state (no change with time) and equilibrium. **Equilibrium** refers to a state in which there is no tendency for spontaneous change. Again, the equilibrium may be dynamic where all net rates of movement equal zero and hence do not change the system's condition. Environmental systems are very rarely at equilibrium because of inputs of heat and energy (even an ice-covered lake is not at equilibrium as the microbial community metabolises food sources accumulated during the ice-free season).

As with equilibrium conditions, environmental systems are seldom at steady state. However, a steady-state approximation is very useful to overcome situations in which you do not have quantitative expressions of system changes over time. Examples of systems in which a steady-state assumption is reasonable include a constant input of chemical to a lake with a constant water residence time, sedimentation rate, etc.; a forest with a constant, average growth rate and water and nutrient inputs; or a population with stable numbers. As Mackay (2001) argues, you can obtain considerable information from a steady-state approximation, including the time response of the system, which is extracted as the global rate constant of the model, e.g. chemical persistence in a system.

Many environmental models must also consider variations in space (e.g. geographical variation) as well as time. If we do not consider spatial variation, then we have a **lumped** model, e.g. $dM/dx = 0$, where $x$ is distance. An example of a lumped model would be treating a lake's water column as a single well-mixed compartment (Figure 6.3.6A). Alternatively, we may consider spatial variation, with the simplest formulation of translating geographical differences into discrete, well-mixed (homogeneous) boxes or compartments, defined according to state and dependent variables. This type of model is often called a **box model** (a highly technical term!). Using the

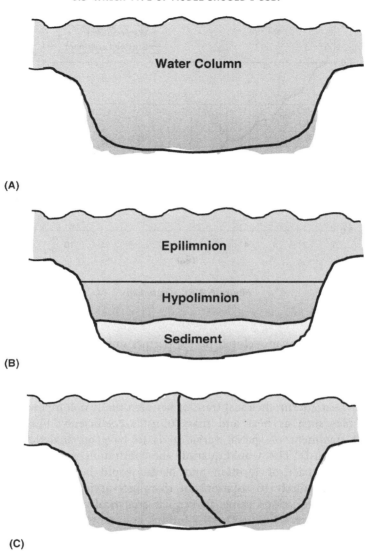

(A)

(B)

(C)

**Figure 6.3.6** Type of models that vary in time and space: (A) a box model with a single, well-mixed (homogeneous) water compartment; (B) a box model that varies in space with two well-mixed water compartments (an upper epilimnion and lower hypolimnion) and a well-mixed sediment compartment; (C) a model system illustrating a continuous change in a water compartment of temperature, oxygen or the concentration of a chemical; (D) output from a time-dependent model; in this case, the model predicts the decline of zinc concentrations in water and sediment in a zinc-contaminated lake. (Adapted from Bhavsar *et al.*, 2004. Dynamic coupled metal TRANSport–SPECiation (TRANSPEC) model: application to assess a zinc contaminated lake. *Environ. Toxicol. Chem.*, **23**, 2410–2420.)

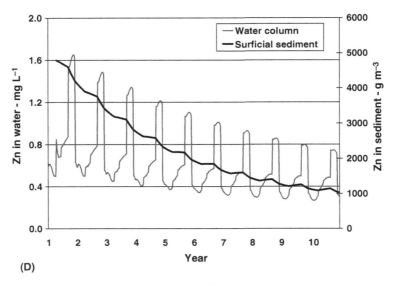

(D)

**Figure 6.3.6**  (*Continued*)

lake as an example, a box model may have a warmer upper water layer or epilimnion and a cooler lower water layer or hypolimnion to address thermal stratification (Figure 6.3.6B). In this example, the **discretisation** is defined according to the **forcing function** (defined below), which in this instance is temperature. The model then includes heat and/or chemical transfer between the two compartments according to variables such as heat and mass transfer coefficients. Finally, the most sophisticated treatment of spatial variation is to have an analytical solution to the differential $dM/dx$. This would quantify the continuous variation in the output as a function of space or location and hence would be a **continuous model** (Figure 6.3.6C). Similarly to solutions of the time-varying differential equation, you can solve the equations using an explicit analytical solution or a numerical approximation (Figure 6.3.6D). These solutions are common among groundwater advection–dispersion models that treat chemical movement in groundwater in (usually) two dimensions. They are less common when considering multiple processes and media because of mathematical complexity.

This brings us to the last group of models – **multi-media models**. Here, the media refer to environmental phases or compartments with differing physical–chemical properties such as air, water, and soil (not a sound and light show). Returning to the lake.... We may assume that it consists of just a well-mixed water column and we treat loss of chemical to the sediments as just that – a loss term. If we want to be more complex, then we would add sediments as another medium since sediments could be a source as well as a sink. For volatile chemicals, we need to consider the air above the water. With our pharmacokinetic model, the media might be blood, lipid, highly perfused tissues, etc.

## 6.4 How do I build a model?

The basic feature of mathematical, 'clear' or 'glass box' models is the quantitative description of the important processes for reproducing the functioning of the environmental system under consideration. In this section, our aim is to define the basic components of this type of model in environmental science, to present the formulations that mathematically depict physical and ecological processes, and to discuss critically the issue of the optimal model complexity that results in reliable answers to the questions posed by the model.

Let us assume that our best knowledge of a hypothetical system is reflected by the conceptual model presented in Figure 6.4.1. In this diagram, the **dependent or state variables** $A$ and $B$ represent components or attributes of the system that we consider relevant to the research question being addressed (this definition is used by ecological modellers whereas engineering modellers use the term **state variable** to refer to the state of the system, such as temperature or pressure, which is considered a **forcing function** by ecological modellers). For example, if we are interested in predicting the success of a strategy to reduce PCB levels in fish from the Baltic Sea, then logical state variables of the model will be the PCB concentrations in the tissues of several fish species. In lake eutrophication models, the state variables can be the various forms of phosphorus (phosphate, dissolved and particulate organic phosphorus) and the different phytoplankton (diatoms, green algae, cyanobacteria) or zooplankton (copepods, cladocerans) groups (Figure 6.4.2). The variables can be expressed in units of mass or concentration (**biogeochemical models**), energy

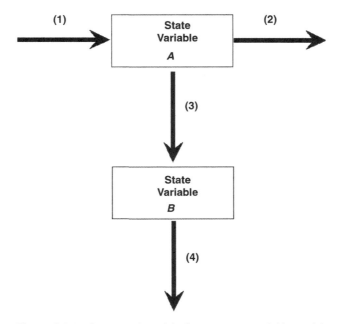

**Figure 6.4.1** Conceptual model of a two-state-variable model.

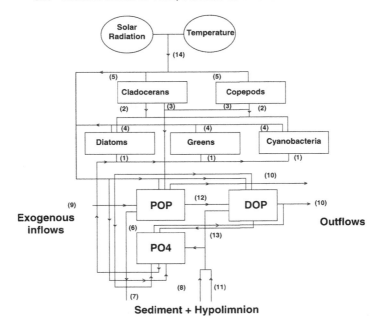

**Figure 6.4.2** The phosphorus cycle as represented by a typical eutrophication model. The simulated processes are: (1) phosphorus uptake from three phytoplankton groups (diatoms, green algae, cyanobacteria); (2) herbivorous grazing from two zooplankton groups (copepods, cladocerans); (3) detrivorous grazing from two zooplankton groups (copepods, cladocerans); (4) phytoplankton basal metabolism excreted as phosphate ($PO_4$), dissolved organic phosphorus (DOP), and particulate organic phosphorus (POP); (5) $PO_4$, DOP, and POP excreted by zooplankton basal metabolism or egested during zooplankton feeding; (6) settling of particulate phosphorus on the sediment; (7) settling of particulate phosphorus in the hypolimnion; (8) water-sediment $PO_4$ and DOP exchanges; (9) exogenous inflows of $PO_4$, DOP, and POP; (10) outflows of $PO_4$, DOP, and POP; (11) fluxes from the hypolimnion; (12) POP hydrolysis; (13) DOP mineralization; and (14) external forcing to phytoplankton growth (solar radiation, temperature). Reproduced by permission of Elsevier from Arhonditsis and Brett (2005a, b). Copyright © 2005, Elsevier. Eutrophication model for Lake Washington (USA): Part II – model calibration and system dynamics analysis. *Ecological Modelling*, **187**, 179–200.

(**bioenergetic models**), and number of species or individuals per unit of volume or area (**biodemographic models**).

The arrow (1) in Figure 6.4.1 corresponds to a variable or function that influences externally the state of the system, and is usually referred to as a **forcing function** of the model. When the model is used as a management tool, we are interested in determining the appropriate levels of certain forcing functions in order to avoid exceeding specific threshold values of the dependent variables. In this context, the model can be used to connect an externally introduced pollutant (e.g. pesticide application or nutrient loading rates) with a key feature of the system (e.g. biodiversity, total phytoplankton or cyanobacteria levels) in order to dictate optimal mitigation strategies. There are also forcing functions that can influence the biotic and abiotic components of the system, although they are not subject directly to anthropogenic control, e.g. temperature and solar radiation (Figure 6.4.2). Generally, as in any open system, we need to specify the prevailing conditions at the system boundary and quantify the constraints imposed upon the various system components.

The arrows (2)–(4) in Figure 6.4.1 represent physical, chemical, and biological processes that transfer mass and/or energy in the system and drive the variability of the state variables. Examples of physical processes are diffusive and advective transport of a fluid such as air or water; chemical reactions that take place in ecosystems such as hydrolysis, photolysis, oxidation, reduction; biological processes of growth, metabolism, mortality, excretion, predation, emigration, or immigration. An interesting feature of mathematical models is the existence of **feedback loops** defined as closed-loop circles of cause and effect in which 'conditions' in one part of the system cause 'results' elsewhere in the system, which in turn amplify (**positive feedbacks**) or counteract (**negative feedbacks**) the original change. In the phosphorus cycle illustrated in Figure 6.4.2, we have the positive feedback of bacteria-mediated mineralisation of the products of the zooplankton basal metabolism that replenishes the summer epilimnetic phosphate pool, which stimulates phytoplankton growth and offsets the herbivorous control of autotroph biomass. Then, this increase of the phytoplankton biomass reinforces the zooplankton growth, and thus prevents an undesirable collapse at the second trophic level. An example of a negative feedback in the system is when excessive phosphorus is added. This causes excessive cyanobacterial growth, which in turn causes shading, which reduces sunlight penetration to lower water depths. This reduces algal growth and the death and sinking of the cyanobacteria causes oxygen consumption and possibly hypolimnetic anoxia, which could kill off benthic organisms.

### 6.4.1 Ecological processes and mathematical equations

A dynamic model of the conceptual diagram in Figure 6.4.1 can be formulated using the **mass conservation principle** as discussed in section 6.3. In quantitative terms, the principle is translated into a **mass balance model** which is built from **mass balance equations** that account for all the inputs and outputs of mass across the system's boundaries and all the transport and transformation processes within the system. For a finite period of time, this concept can be expressed mathematically as:

$$\text{accumulation} = \Delta\text{mass}/\Delta t = \text{volume}\Delta\text{concentration}/\Delta t = \text{input} \pm \text{reactions} - \text{output}$$

Hence

$$V\mathrm{d}A/\mathrm{d}t = \text{process}(1) - \text{process}(2) - \text{process}(3)$$
$$V\mathrm{d}B/\mathrm{d}t = \text{process}(3) - \text{process}(4)$$

where $V$ is volume, $t$ is time, and $A$ and $B$ are concentrations.

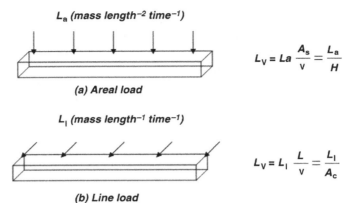

(a) Areal load

(b) Line load

**Figure 6.4.3**  Different ways in which loadings enter water bodies. The symbols are defined as follows: $L_v$, mass per volume and time; $L_a$, mass per area and time; $L_l$, mass per length and time; $L$, total length; $H$, mean depth; $V$, volume; $A_c$, cross-sectional area; and $A_s$, surface area of the water body.

As noted, process (1) acts as a **source** or **input** to the system and may represent human-induced transfer of mass that causes excessive nutrient or pollution problems. The position and the way in which the loading enters an incompletely mixed system are particularly important and determine how it is expressed. For example, some loads from the watershed or the atmosphere enter as a line source along the side of a stream or as an areal source across the top surface area. These inputs are usually expressed as mass per length/area, and to convert them into the proper volumetric units for the mass balance equation we need to divide the loadings by the stream cross-sectional area or the mean depth of the water body (Figure 6.4.3). In contrast, when we model systems that can be considered as completely mixed (e.g. small lakes), all inputs are diluted instantaneously throughout the volume and thus the manner of entry is unimportant. In this case, all loadings can be lumped into a single mathematical expression of time into a single box, i.e. loading $= f(t)$, such as sinusoidal, pulse, step, or linear loading functions (Chapra, 1997).

Process (3) corresponds to the reactions that transform, decompose, or purge nutrients or pollutants from natural systems. These processes are represented typically by first, second or higher order mathematical formulations as in:

$$\text{reaction} = k \times (\text{pollutant mass})^n$$

where $n$ is the order of the reaction, and $k$ is a **rate constant** (e.g. reaction, decay, or excretion) that is usually considered constant for parts or the entirety of the ecosystem. Most of the parameters have physical or ecological interpretation and therefore can be measured in the field or in the laboratory, while many textbooks have compiled information from various sources regarding the plausible ranges of important parameters in environmental sciences (e.g. Jorgensen *et al.*, 1991; Mackay, 2001).

Processes (2) and (4) represent outputs of mass or energy such as the mass carried from the system by an outflow stream or the settling losses of particulate matter across the water–sediment interface. These processes can be modelled as directly proportional to the in-system mass, e.g.

outflow(mass/time) = volumetric flow rate(volume/time)

$\times$ chemical concentration(mass/volume)or settling velocity(distance/time)

$\times$ chemical concentration $\times$ area(distance$^2$)

### 6.4.2  Ordinary and partial differential equations

Thus far, implicit in our discussion is the assumption that the system is spatially homogeneous, i.e. the contents are sufficiently well mixed as to be uniformly distributed. Such a characterisation is often used to model shallow and small lakes where stratification does not occur and horizontal homogeneity can be assumed (Figure 6.4.4a). A common example of this model is the **continuously stirred tank reactor** (CSTR), which simulates the system as a single, well-mixed or homogeneous compartment of volume $V$ where its properties can only vary in time according to the following equation:

$$\frac{\mathrm{d}C}{\mathrm{d}t} = f(C, x, t)$$

where the quantity $C$ (e.g. chemical concentration) being differentiated is the **dependent variable**; the quantity $t$ (time in zero-dimensional systems) with respect to which $C$ is differentiated is called the **independent variable**, and $x$ corresponds to the various inputs of the equation (e.g. external forcing). When the function involves one independent variable the equation is called an **ordinary differential equation** (or ODE). This is in contrast with **partial differential equations** (or PDEs) that involve two or more independent variables. Such equations can be useful for systems with a prevailing one-directional flow, e.g. rivers where the physical, chemical, and biological properties are determined by this flow. In this case, you can simulate the system using a one-dimensional representation that accommodates variability in the $x$ axis (Figure 6.4.4b). An example is the advection–diffusion equation that combines the two main processes of mass transport, advection and diffusion, along with a first-order reaction that describes the spatiotemporal distribution of a substance in a river:

$$\frac{\partial C}{\partial t} = -\frac{\partial Cu}{\partial x} + D_X \frac{\partial^2 C}{\partial x^2} - kC$$

where $C$ is the chemical concentration (mass/volume) in a fixed element of space, $x$ (distance) the direction of the flow, $u$ (distance/time) is velocity for advective transport, $D_x$ (distance$^2$/time) is a diffusion coefficient for diffusive transport, and $k$ (inverse time) is a rate constant for a first-order reaction.

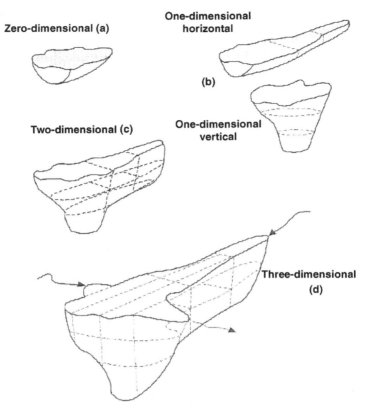

**Figure 6.4.4** Zero-, one-, two-, and three-dimensional approaches for accommodating the spatial variability in environmental systems. Reproduced by permission of Elsevier from Jorgensen, S.E. and Bendoricchio, G. 2001. *Fundamentals of Ecological Modelling*, 3rd edn. Copyright © 2001, Elsevier Amsterdam.

One-dimensional representations can also be used to simulate the vertical stratification of a deep lake without significant variability in the horizontal plane (Figure 6.4.4b). Two or three-dimensional segmentations will be more appropriate for larger systems (estuaries, large lakes with complex morphology, fragmented landscapes) characterised by significant variability of their properties in both horizontal and vertical directions (Figure 6.4.4c and d).

### 6.4.3 What is the most appropriate level of complexity?

Determining the most appropriate level of model complexity is an issue of particular interest to the ecological and environmental science modelling communities (Jorgensen and Bendoricchio, 2001). The selection of a model's complexity entails trade-offs among complexity, information, and uncertainty. Simple models are understood more easily, have fewer parameters unconstrained by the data, and can be subjected more easily to uncertainty analysis, but such minimalistic approaches are often criticised as being crude oversimplifications incapable of reproducing real-world dynamics. Complex models parameterise numerous processes and theoretically have the potential to be more accurately represent complex natural systems. However, when model complexity is not accompanied by sufficient knowledge about the systems being modelled (i.e. in data-poor situations), complex models are criticised for being completely artificial constructs that may include misconceptualisations of the modelled systems. The resulting uncertainty undermines their value.

The problem of complexity was articulated by Costanza and Sklar (1985), who examined how complexity with respect to the number of components, time, and space influenced the effectiveness (the observed variability explained) of 87 mathematical models for freshwater wetlands and shallow water bodies. They were able to show that model effectiveness was positively related with complexity up to a certain level; beyond this level the addition of new state variables and parameters did not improve the model performance (Figure 6.4.5). At a certain level, the knowledge gained from a given model decreased because of the uncertainty underlying the numerous unknown parameters. Based on these results, the authors concluded that the challenge for a modeller is to attain the optimal balance between 'knowing much about little or little about much'.

Arhonditsis and Brett (2004) reported similar results in a meta-analysis of 153 aquatic biogeochemical models published between 1990 and 2002. This study assessed the effects of model complexity (expressed as the number of state variables), spatial dimension (from zero- to three-dimensional models), and simulation period (from

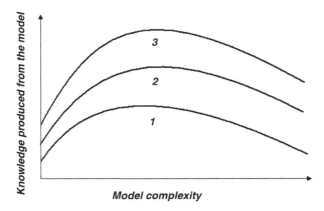

**Figure 6.4.5** The relationship between model complexity and the knowledge gained by the model. The knowledge increases up to a certain level; beyond this level the complexity will not add to our knowledge about the modelled system. At a certain level, the knowledge might even be decreased due to uncertainty underlying the high number of unknown parameters. The line (3) corresponds to an available dataset, which is more comprehensive or has a better quality of data than lines (1) and (2). Reproduced by permission of Elsevier from Jorgensen, S.E. and Bendoricchio, G. 2001. *Fundamentals of Ecological Modelling*, 3rd edn. Copyright © 2001, Elsevier Amsterdam.

days to decades) on model performance. They reported a positive correlation between the number of state variables and the relative error (RE $=\Sigma$ observed values–simulated values/ $\Sigma$ observed values) for phytoplankton ($r = 0.248, p = 0.003$) and zooplankton ($r = 0.626, p < 0.001$) biomass, which shows that a greater number of state variables does not improve model performance. Similarly, a (very weak) positive correlation was found between the duration of the simulation period and the state variable RE values ($r = 0.098, p = 0.022$) indicating that longer simulations resulted in slightly poorer model performance. Marginally significant correlations were also found between the spatial complexity of the models and their (RE values) performance trends ($r = 0.104, p = 0.015$). Their study provided overwhelming evidence that ambitious efforts to increase the level of ecological information mathematically represented by the model, to increase spatial complexity and to use longer simulation periods, do not lead to a systematic improvement in model performance.

With regard to the relationship between knowledge gained and optimum model complexity and structure, Chapra (1997) pointed out that there is a type of a 'Heisenberg uncertainty principle' in ecology that limits our ability to characterise mathematically the complexity of nature. Therefore, some modellers have suggested that we should opt for the simplest model in the context being developed, i.e. we should start with simple approaches and proceed to greater complexity as warranted by data availability, the system being studied, and the questions being asked (Franks, 1995; Mackay, 2001). This notion is often referred to as applying Ockham's razor in order to optimise the number of assumptions or unknown model nodes (e.g. state variables, parameters) given our knowledge of the modelled system. In the words of Einstein, 'every model should be as simple as possible, but no simpler'.

## 6.5  Steps in developing a model

In this section we elaborate the different steps involved during the model development process. We emphasise the special practices used during the calibration, predictive evaluation, and uncertainty analysis of a mathematical model. We conclude with a critical discussion of the feasibility of complete confirmation and what should be considered as 'proper' evaluation of a mathematical model.

Most of the modelling textbooks present graphs similar to the one shown in Figure 6.5.1, which illustrates the process of model development. Most modellers agree that the first step in the modelling procedure should be the definition of the problem, i.e. before you start writing equations or learning modelling software, you must clearly specify the problem, the objectives of the model, and define system boundaries. In this phase, the modeller should consult all the existing information pertaining to the study system, i.e. data related to the physical, chemical, and biological properties of the system, and possibly socio-economic information, and the historical and legislative context. By being articulate at this stage, you minimise the likelihood of being diverted into irrelevant activities and misallocating your precious time!

After you specify the problem being addressed, the next step is to develop a clear conceptual model that defines and relates explicitly components of the model. What

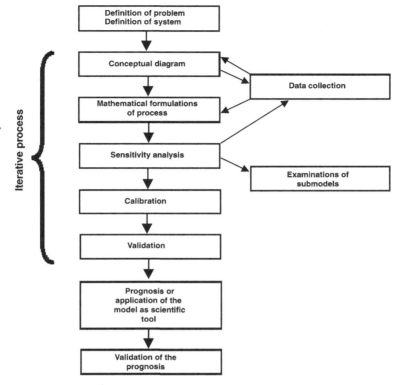

**Figure 6.5.1**  A tentative procedure that should be followed during the development process of a model. Note the iterative character of this protocol, which corresponds to the notion that the model should evolve in a parallel manner with our knowledge of the ecosystem functioning. Reproduced by permission of Elsevier from Jorgensen, S.E. and Bendoricchio, G. 2001. *Fundamentals of Ecological Modelling*, 3rd edn. Copyright © 2001, Elsevier Amsterdam.

is the essential number of components I need to include? What processes or mechanisms relate the components? Are there feedback loops? What characteristics of the system appear to be critical and hence demand more attention? What are my simplifying assumptions?

As the model begins to take shape, you must delineate the optimal spatiotemporal resolution and the sub-models required to attain the goals stated previously. Typical questions posed in relation to the model characterisation are:

'What is the most suitable model segmentation to accommodate the spatial variability of the system?'
'What is the longest period for which the model can provide realistic estimates or predictions?'
'Which chemicals and mechanisms need to be considered for effectively simulating system perturbation?'

Given the high number of the 'known or unknown unknowns' in a mathematical model, we recommend avoiding final decisions with regards to the model structure at this stage. Rather, the model could be used to direct collecting additional data, and if you cannot get more data, then the original structure should be trimmed down to avoid significant model uncertainty due to missing information. While ideal, this preliminary application of a model to determine the missing information from the system is reported rarely in the modelling literature. In real life, the models are developed after the collection of the data and their structure is a compromise between the objectives of the model and the available information.

In the next phase, the modeller aims to describe the selected physical or ecological processes with a system of mathematical equations, which then will be implemented on the computer. You will either code this yourself or find an existing model that meets your needs. If coding the model yourself, you need to make technical decisions such as the selection of the numerical methods for solving the ODEs or PDEs, choosing a computer language, debugging the code, and finally testing and evaluating the model. There are several excellent presentations of these topics in the modelling literature (Chapra, 1997; Chapra and Canale, 1998; Jorgensen and Bendorrichio, 2001). We focus on three steps of the model development process (**calibration**, **predictive evaluation**, and **sensitivity/uncertainty** analysis) that are critical in constructing a credible model in environmental science.

**Sensitivity analysis** is an important step that facilitates insight into the behaviour of the model. Specifically, with this process the modeller evaluates the model sensitivity to the parameters, forcing functions, or state-variable sub-models. One common way for carrying out sensitivity analysis is the simple perturbation of each element of the input vector, and the estimation of the corresponding response on selected state variables. Thus, the sensitivity $S$ of a state variable $y$ to variations in a parameter $\theta$ is defined as follows:

$$S = (\partial y / y)/(\partial \theta / \theta)$$

The magnitude of the perturbation is usually based on our knowledge of the uncertainty of the parameters: we advise assessing the sensitivity of a state variable at two or more levels of a parameter's relative change (e.g. $\pm 10\%$, $\pm 20\%$, and $\pm 50\%$), as the relationship between a parameter and a state variable can be nonlinear. A sensitivity analysis of the process equations can suggest structural changes in the model. Most of the well-studied ecological processes can be described mathematically by a variety of relationships that entail different assumptions and complexity levels, e.g. Monod and Variable-Internal Stores (VIS) models for modelling the phytoplankton uptake of nutrients from the water column and their conversion into biomass. Thus, a modeller can examine the effects of the two different formulations on the patterns of key state variables (e.g. phytoplankton, dissolved oxygen), and then decide which is the most appropriate sub-model in the context of the questions being addressed. A sensitivity analysis of forcing functions illuminates the effects of the external forcing on the model outputs, and indicates the level of accuracy required for the forcing function data (e.g. solar radiation, temperature, chemical loading) if the model is to produce meaningful output.

Commonly modellers are confronted with what is called an **inverse** problem: there is sufficient information on the levels and the variability of the state (dependent) variables, but little is known about the values of the model variables or parameters. The procedure by which the modeller adjusts the model variables to find the best agreement between modelled and measured or observed data is called **calibration**. In essence, our goal with calibration is to improve the estimation of the independent variables or parameters of the model. Our efficiency during the calibration task can be improved significantly if we are able to narrow down the number of model solutions that need to be examined by assigning realistic ranges of plausible values to each variable. For example, if we have a model with 10 variables and the under-lying uncertainty requires the testing of 10 values for each variable, then to attain an acceptable fit to the data we need to run the model $10^{10}$ times. However, if based on the literature information the number of plausible values for each variable is only five, the calibration of the model needs $10^5$ runs, which entails significantly fewer computational demands. During the calibration of the model, the modeller should keep in mind that: (1) all models in ecology and environmental sciences are drastic simplifications of reality that merely approximate the actual processes, because individual equations that are approximately correct in controlled laboratory envir-onments may not collectively yield an accurate picture of the ecosystem functioning; and (2) all parameters are effectively spatially and temporally averaged in some scale, lumped across an inconceivably heterogeneous environment and/or biological com-munity to reproduce average values that are unlikely to be represented by a fixed constant. Hence, the obtained values of variables usually differ from the real (but unknown) ones, and this discrepancy reflects our insufficient knowledge of the real system dynamics along with all the assumptions and approximations being made when we set up our mathematical model.

Model calibration can be carried out by trial and error or by using *optimisation* techniques. The advantage of the first approach is that it allows the modeller to become acquainted with the model's responses to changes in the different parameters. On the other hand, optimisation methods are designed to search the parameter space for combinations of values of variables which provide the best fit through minimising **cost** or **objective functions** (i.e. functions that measure the discrepancy between observed data and model outputs). Therefore the latter strategy ensures that the calibration parameter is optimised and that a significant lack of fit is due to the inadequacy of the model structure and not due to poor choice of the value of that variable. There are many techniques designed to solve optimisation problems: (1) **one-dimensional optimisation** methods involving functions that depend on a single variable (Golden-section search, quadratic interpolation, Newton's method); (2) **multi-dimensional optimisation**: (a) **direct methods** – random (simulated annealing, genetic algorithms, artificial neural networks), univariate, and pattern searches (Powell's method) that do not require the evaluation of the objective function's derivatives; (b) **gradient methods** that do require the evaluation of the function's first- or second-order derivatives such as the steepest ascent/descent, conjugate gradient (Fletcher-Reeves), and Marquardt's methods; and (3) **constrained optimisation** methods that include linear or non-linear, equality or inequality constraints (Simplex method, Generalised reduced gradient search). Despite the conceptual and practical advantages of model optimisation, Arhonditsis and Brett (2004) found that only a small proportion (8.5%) of the aquatic biogeochemical modelling studies base their calibration results on optimisation methods. For some inexplicable reason, modellers seem reluctant to embrace such techniques and include them in their repertoire when developing models in environmental sciences.

Throughout the calibration process, a dynamic or time-dependent model should be run with the same external forcing for a sufficient length of time to ensure that it is stabilised, i.e. that the same pattern is repeated (Figure 6.5.2). The outputs from the so-called 'equilibrium' phase should be compared against the observed data and then it should be decided whether or not a satisfactory calibration of the model is achieved. There is an extensive literature on the appropriate measures of fit for objectively evaluating the performance of a mathematical model (Reckhow *et al.*, 1990; Mayer and Butler, 1993; Power, 1993). Here, we present selectively five of the most frequently used model fit statistics:

(a) Correlation coefficient ($r$) between model predictions and observations:

$$r = \frac{\sum_{i=1}^{n}(O_i - \bar{O})(M_i - \bar{M})}{\sqrt{\sum_{i=1}^{n}(O_i - \bar{O})^2 \sum_{i=1}^{n}(M_i - \bar{M})^2}}$$

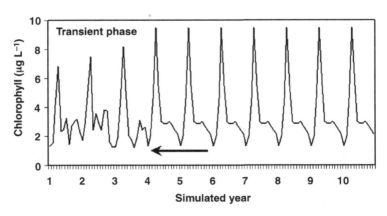

**Figure 6.5.2** Transient and equilibrium phase of a mathematical model that is run with the same annual external forcing for 10 years. The first 3 years correspond to the transient phase when the model is dependent on the initial conditions. After the third year (black arrow), the model enters the 'equilibrium' phase when the same pattern is repeated each year.

where $n$ is the number of observations, $O_i$ is the $i$th of $n$ observations, $M_i$ is the $i$th of $n$ predictions, and $\bar{O}$ and $\bar{M}$ are the observed and predicted averages, respectively. The correlation coefficient $r$ measures the tendency of the predicted and observed values to vary together linearly. It can range from $-1$ to $1$; correlation values close to 1 do not indicate that the predicted and observed values match each other but only that they tend to vary in a similar way. Negative values indicate that the observed and predicted values tend to vary inversely.

(b) Average error (AE)

$$AE = \frac{\sum_{i=1}^{n}(M_i - O_i)}{n} = \bar{M} - \bar{O}$$

The average error is a measure of the size of the discrepancies between predicted and observed values. Values near zero indicate a close match. The average error is a measure of aggregate model bias, although values near zero can be misleading because negative and positive discrepancies can cancel each other.

(c) Root mean squared error (RMSE):

$$RMSE = \sqrt{\frac{\sum_{i=1}^{n}(M_i - O_i)^2}{n}}$$

The root mean squared error is another measure of the size of the discrepancies between predicted outputs and observed values. The root mean squared error accommodates the shortcoming of the average error by considering the magnitude rather than the direction of each discrepancy. Together the two statistics provide an indication of model prediction accuracy.

(d) Reliability index (RI)

$$RI = \exp\sqrt{\frac{1}{n}\sum_{i=1}^{n}\left(\log\frac{O_i}{M_i}\right)^2}$$

The reliability index (Leggett and Williams, 1981) quantifies the average factor by which model predictions differ from observations. An RI of 2.0 indicates that a model predicts the observations within a multiplicative factor of two, on average. Ideally, the RI should be close to 1. When the root mean squared error is calculated for log-transformed values of the predictions and observations, then the RI is the exponentiated RMSE.

(e) Modelling efficiency (MEF)

$$MEF = \frac{\left( \sum_{i=1}^{n} (O_i - \bar{O})^2 - \sum_{i=1}^{n} (M_i - O_i)^2 \right)}{\sum_{i=1}^{n} (O_i - \bar{O})^2}$$

The modelling efficiency measures how well a model predicts relative to the average of the observations. It is related to the RMSE according to $MEF = 1 - RMSE^2/s^2$ where $s^2$ is the variance of the observations. A value near 1 indicates a close match between observations and model predictions. A value of zero indicates that the model predicts individual observations no better than the average of the observations. Values less than zero indicate that the observation average would be a better predictor than the model results. Performance tests should not neglect internal rates and derived quantities from the model (e.g. primary production, hydrolysis, particulate matter deposition, $f$ ratios) when testing whether the simulation reproduces realistically the functional properties of the system modelled. In cases where relevant data are not available, it should be indicated clearly that the internal structure of the model could not be tested, and therefore the match between state variables and observed data might not be the result of a correct model solution, i.e. we obtain 'good results for the wrong reasons'.

As the modeller becomes acquainted with the model, especially if a trial and error calibration scheme is used, he/she realises that several distinct choices of model inputs lead to the same model outputs, i.e. many sets of parameters fit the data about equally well. This non-uniqueness of the model solutions characterises what is known in the modelling literature as **equifinality** (Beven and Binley, 1992). A main reason for the equifinality (poor identifiability) problem is that the causal mechanisms/hypotheses used for understanding how the system works internally is of substantially higher order than what can be observed externally (Beck, 1987). However, determining a model structure (and associated parameter values) that reflects realistically the natural system dynamics is particularly important when we aim to make predictions for future hypothesized states. For example, when a water quality model does not operate with realistic ecological structure (e.g. relative/absolute magnitudes of biological rates and transport processes), even if the fit between model outputs and observational data is satisfactory, its credibility to provide predictions about how the system will respond under significantly different external nutrient loading conditions is very limited.

The recognition of the equifinality problem is progressively changing the perspective of modellers from seeking a single 'optimal' value for each model parameter, to seeking a **distribution of parameter sets** that all meet a pre-defined fitting criterion (Figure 6.5.3). These acceptable parameter sets may then provide the basis for

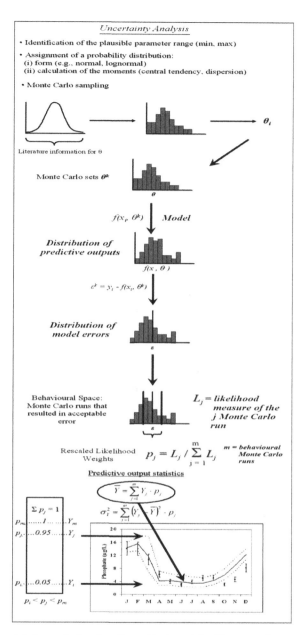

**Figure 6.5.3** Uncertainty analysis of mathematical models. The procedure scrutinizes the parameter space and selects parameter sets that result in acceptable error. The Generalized Likelihood Uncertainty Estimation method assigns likelihood weights to the selected parameter sets, as derived from measures of fit (or likelihood measures). These weights reflect the ability of the different parameter sets to reproduce acceptably 'non-error-free' observations from the environmental system. The weighting assigned to the retained behavioural runs is propagated to the model output and forms a likelihood-weighted cumulative distribution of the predicted variable(s), which are then used for estimating the prediction uncertainty ranges.

estimating model prediction error associated with the model parameters. The development of methods for identifying plausible parameter sets for large multi-parameter environmental models with limited observational data began with the work of Hornberger and Spear (1981). Their method, called regionalised (or generalised) sensitivity analysis (RSA), is a Monte Carlo sampling approach to assess model parameter sensitivity. Regional sensitivity analysis is simple in concept, and is a useful way to use limited information to bound model parameter distributions. Given a particular model and an environmental system being modelled, the modeller first defines the plausible range of certain key model response variables as the 'behaviour'. For example, based on existing information from the system the chlorophyll $a$ concentration lies within the range of 5–25 µg L$^{-1}$. The modeller then samples from (usually) uniform distributions of each of the model parameters and computes the chlorophyll $a$ values. All parameter sets that result in chlorophyll $a$ predictions within the 'behaviour' range are termed 'behaviour generating' and thus become part of the model parameter distribution. The parameter sets that do not meet this behaviour criterion are termed 'non-behaviour generating'.

Generalised Likelihood Uncertainty Estimation (GLUE) is another popular parameter estimation method, which is an extension of the binary 'acceptance/rejection' system of behavioural/non-behavioural simulations of the original Regionalised Sensitivity Analysis. The GLUE method uses measures of fit (or likelihood measures) to assign different levels of confidence (weighting) to different parameter sets, reflecting their ability to reproduce acceptably 'non-error-free' observations from the environmental system (Beven and Binley, 1992). The GLUE procedure also requires a large number of Monte Carlo model runs sampled from distributions across a plausible range of each parameter. The behavioural runs are selected on the basis of a subjectively chosen threshold of the likelihood measure (e.g. RMSE for total phosphorus $\leq 5$ µg L$^{-1}$) and are rescaled so that their cumulative total is 1.0. The weighting assigned to the retained behavioural runs is propagated to the model output and forms a likelihood-weighted cumulative distribution of the predicted variable(s), which are then used for estimating the prediction uncertainty ranges.

We emphasise that model calibration does not provide any information with regard to the model's predictive ability; it merely shows that the model can 'fit' a single dataset. Therefore, calibration should always be followed by a procedure whereby the modeller tests the model against an independent set of data in order to observe how well the model fits these data. This phase is usually referred to as model **predictive evaluation**. The literature often refers to this phase as 'validation'. However, as Oreskes *et al.* (1994) argue, environmental models that necessarily deal with open systems and myriad sources of variability and uncertainty cannot be truly validated.

A common practice during the predictive evaluation of models is to split the dataset into two subsets, and use the first subset for calibration and the second for evaluating its predictive ability. Although this procedure can be insightful, the evaluation rigour depends on the difference between the calibration and evaluation datasets. For example, in studies intended to model system responses to increasing chemical loading, the evaluation dataset should describe conditions that differ significantly from those used during the calibration process. In cases where such datasets are not available and the two subsets (calibration and predictive evaluation) are essentially identical, the modellers should be aware that they showed mainly the model's ability to run in a stable mode and have not tested unequivocally its capacity to predict different conditions.

If, during evaluation, model predictions do not match the evaluation set of observed data, then this might be a sign that the model structure should be modified by including additional mechanisms, considering alternative formulations, or refining the model structure. On the other hand, even if the model fits the data, the modeller should acknowledge that he/she simply proved the model's predictive ability for the range of conditions defined by the original calibration and the evaluation datasets. Model acceptance in two or more case studies is not evidence for a general statement, but merely the start of a 'perpetual race' for confirmation (Oreskes *et al.*, 1994). In fact, the 'truth' of a model can never really be proven absolutely, it can only be supported by the range of conditions examined and its robustness is still a matter under investigation. The greater the number of settings for which the model is tested and confirmed, the higher the likelihood that its structure and conceptualisations are not fundamentally defective. As put by Oreskes *et al.* (1994), 'confirming observations do not demonstrate the veracity of a model or hypothesis; they only support its probability'.

The modelling procedure presented in this section has been recommended strongly in the literature, and several modellers have argued emphatically that there is a need for methodological consistency. Nonetheless, a recent meta-analysis showed that at least in the field of aquatic mechanistic biogeochemical modelling the large majority of the published studies over the past decade did not assess prediction error properly. Aquatic mechanistic modellers are still reluctant to embrace uncertainty analysis techniques and assess the reliability of the critical planning information generated by the models (Arhonditsis and Brett, 2004). Only 30.1% of aquatic biogeochemical modelling studies published during the past decade evaluated statistically the model performance; thorough quantification of model sensitivity to parameters, forcing functions, and state variable sub-models was only reported in 27.5% of the studies, while 45.1% of the published models did not report any results of uncertainty/ sensitivity analysis. After several decades of experience and many convincing presentations in several classic modelling textbooks of what 'rational model development' is (e.g. Chapra, 1997; Jorgensen and Bendoricchio, 2001), the absence of a systematic methodological protocol widely followed by the modellers is surprising. We should understand that the methodological consistency in the model development process is an analogue to the way a chemical analyst strives to attain clean laboratory conditions, excellent calibration plots, and faithful adherence to the analytical protocol.

Regardless of which model you use or develop, we hope the process of model development is enlightening – that you learn about the data, the system, and how the system works.

## 6.6  Illustrative case study

To illustrate the principles and procedures of model development and testing introduced and discussed earlier in this chapter, Box 6.6.1 presents, as a case study, a complex eutrophication model that has been developed as a management tool for Lake Washington, USA.

---

### 6.6.1  Eutrophication model

#### Definition of the problem

Lake Washington is the second largest natural lake in the State of Washington, with a surface area of 87.6 km$^2$ and a total volume of 2.9 km$^3$. The mean depth of the lake is 32.9 m (maximum depth 65.2 m), the summer epilimnion depth is typically 10 m and the epilimnion: hypolimnion volume ratio during the summer is 0.39. The retention time of the lake is on average 2.4 years. Lake Washington has been studied extensively and it is perhaps the best example in the world of successful lake restoration by sewage diversion (Edmondson, 1994). Because of the lake's proximity to the Seattle metropolitan area, severe eutrophication, cyanobacteria dominance, and declining water quality resulted from increasing amounts of secondary treated sewage between 1941 and 1963. Sewage was diverted from the lake between 1963 and 1967, with discharge of wastewater treatment plant effluents (except for combined sewer overflows) eliminated by 1968. Rapid water quality improvements followed along with a dramatic decline of cyanobacteria abundance. Currently, Lake Washington can be characterised as a mesotrophic ecosystem with limnological processes strongly dominated by a recurrent diatom bloom, with epilimnetic chlorophyll concentration peaks on average at 10 µg L$^{-1}$, which is 3.2 times higher than the concentrations during the summer stratified period (Arhonditsis et al., 2003). However, the increasing urbanisation in the watershed along with ominous projections of the city's population may again pose threats to the sustainability of the present quality status of the lake. Thus, the model described here aimed to evaluate alternative management schemes and assess the lake's response to increasing external nutrient loading conditions.

#### Model description

The physical structure of the model is simple and consists of two spatial compartments representing the lake epilimnion and hypolimnion. The depths of the two segments vary with time and are defined explicitly based on extensive field measurements for the study period 1994–2000 (Figure 1). The ecological sub-model consists of 28 state

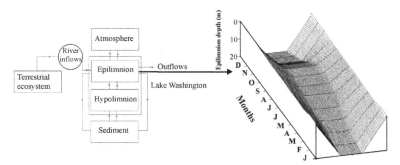

**Figure 1** The flow diagram of the Lake Washington eutrophication model, consisting of two spatial compartments (epilimnion and hypolimnion). The right plot depicts the annual variability of the epilimnion compartment. Reproduced by permission of Elsevier from Arhonditsis and Brett (2005a, b). Copyright © 2005, Elsevier.

variables and simulates five elemental cycles (organic C, N, P, Si, O) as well as three phytoplankton (diatoms, green algae, and cyanobacteria) and two zooplankton (copepods and cladocerans) groups. The forcing functions of the model were: epilimnetic and hypolimnetic temperature, solar radiation, precipitation, evaporation, river inflows, and outflows from the system. Field measurements and meteorological data from the Seattle area were used to reproduce the mean annual hydrological and nutrient loading cycle of the lake.

The three phytoplankton functional groups differ with regard to their strategies for resource competition (nitrogen, phosphorus, light, temperature) and metabolic rates, as well as their morphological features (settling velocity, shading effects). Diatoms are modelled as r-selected organisms with high maximum growth rates and higher metabolic losses, strong phosphorus and weak nitrogen competitors, lower tolerance to low light availability, low temperature optima, silica requirements, and high sinking velocities. By contrast, cyanobacteria are modelled as K-strategists with low maximum growth and metabolic rates, weak P and strong N competitors, higher tolerance to low light availability, low settling velocities, high temperature optima, and higher shading effects (i.e. filamentous cyanobacteria). The parameterisation of the third functional group (labelled as 'green algae') aimed to provide an intermediate competitor and depict more realistically the continuum between diatom- and cyanobacteria-dominated phytoplankton communities in Lake Washington. The herbivorous zooplankton community consists of two functional groups, i.e. cladocerans and copepods, and their biomass is controlled by growth, basal metabolism, higher predation, and outflow losses. The zooplankton grazing term considers explicitly algal food quality (fatty acid, amino acid, protein content, and/or digestibility) effects on zooplankton assimilation efficiency, and also takes into account recent advances in stoichiometric nutrient recycling theory (Arhonditsis and Brett, 2005a). The two herbivores modelled differ with regard to their grazing rates, food preferences, selectivity strategies, elemental somatic ratios, vulnerability to predators, and temperature requirements (Arhonditsis and Brett, 2005a,b). These differences drive their successional patterns along with their interactions with the phytoplankton community. The specific parameterisation assigned to cladocerans aimed to more closely represent *Daphnia* dynamics, which is the dominant member of the Lake Washington zooplankton community.

The model also considers another critical component for predicting a lake's response to different nutrient loading conditions, i.e. sediment–water interactions. Because of the limited amount of information from the system, the sediment sub-model is a simple dynamic approach that relates sediment oxygen consumption, nitrogen, and phosphorus release to the sedimentation and burial rates. The relative magnitudes of ammonium and nitrate fluxes are determined by the nitrification occurring at the sediment surface. The parameter values for the sediment sub-model were assigned prior to model calibration and were based on estimates from nutrient budget calculations and some field measurements that cover a wide range of nutrient loading in Lake Washington (pre-diversion period, transient phase, and current conditions), which provided reasonable accuracy in approximating sediment response; at least, at the level of net total annual sediment fluxes.

## Sensitivity analysis

The sensitivity analysis of the model involved several steps in which the role of the different model inputs (parameters, forcing functions) was assessed. Each of the model parameters was assigned a prior probability distribution formulated from field observations, laboratory studies, literature information, and expert judgment. These distributions reflected the relative plausibility of the different parameter values and were sampled independently to generate $10^5$ input vectors. These parameter sets were used to run the model and identify the most influential model parameters on the spatiotemporal outputs of the model. This part of the analysis showed that the zooplankton maximum grazing rate, the phytoplankton basal metabolism, the phytoplankton maximum growth rate, the zooplankton grazing half saturation constant, and the background light attenuation exert the greatest impact on model behaviour. The second phase examined the role of plankton stoichiometries and identified the maximum phosphorus uptake rate and the phosphorus intracellular storage capacity as the most influential parameters on key model endpoints (epilimnetic and hypolimnetic phytoplankton biomass, proportion of cyanobacteria, and total zooplankton biomass). Finally, the evaluation of the role of the forcing function uncertainties highlighted the importance of having accurate nutrient loading estimates for simulating the dynamics of the system. All these results were used to guide the calibration of the model and to reproduce the Lake Washington patterns.

# Model calibration

The data-set used for model calibration consisted of observed data collected from twelve inshore and offshore sampling stations between January 1995 and December 2001. The eutrophication model was calibrated by tuning the model parameters within their observed literature ranges. The comparison between simulated and observed phosphate, total phosphorus, dissolved oxygen, and plankton abundance values in Lake Washington's epilimnion and hypolimnion is shown in Figure 2.

In these plots, the dots correspond to the mean volume weighted averages, while the error bars represent the standard deviations for the monthly parameter values. It can be seen that the model provides a reasonable reproduction of the mean seasonal epilimnetic patterns for zooplankton biomass, chlorophyll $a$, phosphate, and total phosphorus concentrations. The predicted hypolimnetic annual chlorophyll cycle is in agreement with the observed data, with the only exception being the late spring (May–June) levels when the model predicts a more rapid phytoplankton biomass decline. Total phosphorus dynamics in the hypolimnion are simulated accurately, whereas the model predicts a greater hypolimnetic phosphate ($\approx 20\,\mu g\,L^{-1}$) accumulation during the stratified period.

The simulated plankton succession patterns are as follows (Figure 3): the diatoms are the dominant phytoplankton group (55%) during the spring bloom, when green algae and cyanobacteria account for 33% and 18% of the total phytoplankton biomass, respectively. In contrast, the summer phytoplankton community is divided almost equally among these three groups. Both results are very close to the current Lake Washington phytoplankton seasonal succession patterns (Arhonditsis *et al.*, 2003). The model also describes accurately the zooplankton annual cycle, with copepod dominance throughout the winter until the end of April or mid-May and cladoreran dominance during the summer and autumn period.

The assessment of the goodness-of-fit between the simulated and observed monthly values (average error (AE), relative error (RE), coefficient of determination ($r^2$)) for the Lake Washington epilimnion, hypolimnion and overall lake volume is presented in Table 1. Generally, a fairly low bias and high accuracy characterises the model patterns for zooplankton, dissolved oxygen, total nitrogen, total phosphorus, phosphate, and silica concentrations. The lower $r^2$ values for the hypolimnion are largely a statistical artefact, because there is much less annual variability to be explained in these data. This is also true for the total organic carbon (TOC) concentrations ($r^2 < 0.10$), since both the AE ($-0.30-0.08$ mg $L^{-1}$) and RE (11–20%) values are satisfactory. A good agreement is obtained between the simulated and observed epilimnetic nitrate concentrations (AE $= 0.40$ µg $L^{-1}$, RE $= 11\%$, $r^2 = 0.95$), which decreases in the hypolimnion (AE $= 95.33$ µg $L^{-1}$, RE $= 36\%$, $r^2 = 0.43$), and the overall lake results (AE $= 33.98$ µ g $L^{-1}$, RE $= 16\%$, $r^2 = 0.89$). Finally, ammonium is the state variable with the poorest performance with fairly high AE ($>15$ µ g $L^{-1}$) and RE ($>65\%$) and low $r^2$ ($<0.10$) values.

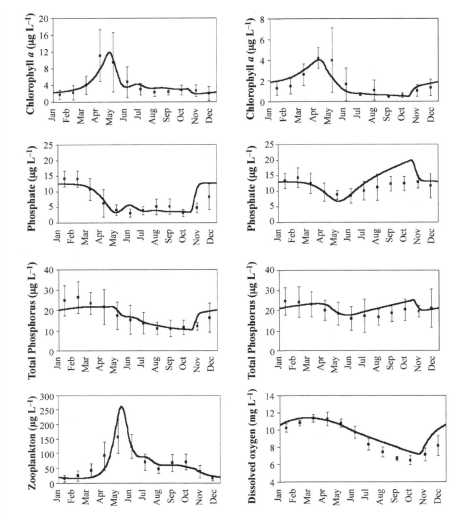

**Figure 2** Comparison between simulated (black line) and observed values for Lake Washington: (left panel) epilimnetic chlorophyll *a*, phosphate, total phosphorus, zooplankton; (right panel) hypolimnetic chlorophyll *a*, phosphate, total phosphorus, and dissolved oxygen. The dots correspond to the mean volume weighted averages, while the error bars represent the standard deviations for the monthly observed values for all stations and years (1995–2001) in the King County monitoring programme (Arhonditsis *et al.*, 2003). Reproduced by permission of Elsevier from Arhonditsis and Brett (2005a, b). Copyright © 2005, Elsevier.

## Model predictive evaluation

Lake Washington provides an excellent case for testing whether the model can predict responses to variations in external forcing, as the existing data span a

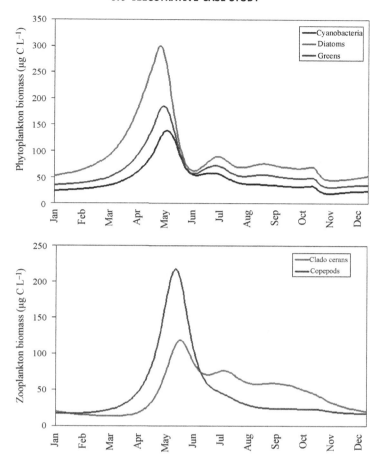

**Figure 3** Seasonal phytoplankton and zooplankton succession patterns as simulated by the model. Reproduced by permission of Elsevier from Arhonditsis and Brett (2005a, b). Copyright © 2005, Elsevier.

40-year period with a wide range of nutrient loading conditions, i.e. the period when the wastewater effluents were diverted from the lake (1963–1968), the transient phase (1969–1974), and current conditions (Edmondson, 1997). Among the variety of predictive evaluation tests presented in Arhonditsis and Brett (2005b), an interesting assessment was the performance of the model relative to the 1962 nutrient loading; the year when Lake Washington received the maximum input of treated sewage effluent. For this predictive evaluation test, it is important to note that the structure of the eutrophication model is not compatible directly with the pre-diversion years, because during that period *Daphnia* was not a prominent component of the zooplankton community. *Daphnia* became established after 1976 as a result of factors that were not accounted for by the model, i.e. bottom-up (due to the disappearance of the filamentous cyanobacterium *Oscillatoria*) and

**Table 1** Goodness-of-fit statistics for the Lake Washington eutrophication model, based on the monthly values (1995–2001) of dissolved oxygen (DO), total nitrogen (TN), nitrate ($NO_3^-$), ammonium ($NH_4^+$), total phosphorus (TP), phosphate ($PO_4^{3-}$), chlorophyll $a$ (chl$a$), total organic carbon (TOC), total zooplankton biomass (ZOOP), and total epilimnetic silica (Si)

| | Statistic | DO (mg L⁻¹) | TN (µg L⁻¹) | $NO_3^-$ (µg L⁻¹) | $NH_4^+$ (µg L⁻¹) | TP (µg L⁻¹) | $PO_4^{3-}$ (µg L⁻¹) | chl$a$ (µg L⁻¹) | TOC (mg L⁻¹) | ZOOP (µg L⁻¹) | Si (mg L⁻¹) |
|---|---|---|---|---|---|---|---|---|---|---|---|
| Epilimnion | Average error | −0.09 | −12.10 | 0.40 | 19.74 | −0.10 | −0.53 | 0.01 | 0.08 | −1.61 | −0.07 |
| | Relative error (%) | 5 | 7 | 11 | 84 | 12 | 23 | 20 | 11 | 25 | 27 |
| | $r^2$ | 0.77 | 0.90 | 0.95 | 0.01 | 0.79 | 0.71 | 0.91 | 0.10 | 0.88 | 0.49 |
| Hypolimnion | Average error | −1.23 | 46.67 | 95.33 | 22.12 | 2.76 | 4.06 | −2.32 | −0.30 | | |
| | Relative error (%) | 14 | 15 | 36 | 85 | 17 | 37 | 137 | 21 | | |
| | $r^2$ | 0.91 | 0.72 | 0.43 | 0.10 | 0.32 | 0.43 | 0.69 | 0.00 | | |
| Water column[a] | Average error | −0.47 | 3.39 | 33.98 | 15.65 | 0.03 | 0.07 | −0.38 | −0.03 | | |
| | Relative error (%) | 7 | 5 | 16 | 65 | 8 | 15 | 22 | 12 | | |
| | $r^2$ | 0.86 | 0.85 | 0.89 | 0.07 | 0.79 | 0.68 | 0.93 | 0.02 | | |

[a] Averages weighted over the epilimnion and hypolimnion volume.

top-down (due to reduced predation by mysid shrimp) control of the zooplankton community (Edmondson, 1994). These structural changes in the Lake Washington food web indicate that the models cannot replicate the enormous complexity of natural systems, and thus the modelling procedure should be considered as an iterative process where model formulation and predictive evaluation criteria always evolve in a parallel manner with the real system (Rykiel, 1996). To attain a plankton community similar to the one residing in Lake Washington during the pre-diversion period, the zooplankton group labelled as 'Cladocerans' was inactivated and then the model was run using the 1962 nutrient loading (Figure 4). The model predicted a chlorophyll annual cycle with a maximum of $38 \mu g L^{-1}$ and mean summer concentrations of $10 \mu g L^{-1}$. In addition, cyanobacteria became the predominant phytoplankton group and their proportion varied between 30 and 95% of the total phytoplankton during the year. Total nitrogen, total phosphorus, and phosphate levels were also fairly close to those reported in the literature (Edmondson and Lehman, 1981).

## Scenarios analysis

The eutrophication model was also used to examine two alternative management scenarios that involved an increase in the nutrient loading equivalent to 25% of the 1962 point-source loading levels. In the first case, the sewage effluents were discharged exclusively into the hypolimnion (scenario 1), and in the second scenario effluents were discharged equally between the epilimnion and hypolimnion. In the first loading scenario, most of the incoming nutrients would not be directly available for phytoplankton uptake until autumn mixing when physical conditions (light, temperature) are not favourable for phytoplankton growth. The second scenario considers nutrient supply to the epilimnion during the summer-stratified period, when the phytoplankton is phosphorus limited and more susceptible to cyanobacteria dominance. The system dynamics were studied through a variety of meteorological conditions (temperature, solar radiation, precipitation). Based on individual month variability, 150 different annual cycles of meteorological forcing were generated and used for identifying structural shifts in the lake patterns between the current state and the two nutrient loading scenarios.

Summary statistics for the annual values for chlorophyll $a$, total nitrogen (TN), nitrate ($NO_3^-$), total phosphorus (TP), and phosphate ($PO_4^{3-}$) concentrations, and the proportion of cyanobacteria are presented in Table 2. Both nutrient loading scenarios resulted in an increase of the mean annual phytoplankton biomass of about $1 \mu g$ chl$a L^{-1}$, and an epilimnetic cyanobacteria proportion that was 6–7% higher than the current level. In addition, there were simulated years for the second scenario (effluents distributed between epilimnion and hypolimnion), when epilimnetic cyanobacteria comprised an average of 38–40% of the

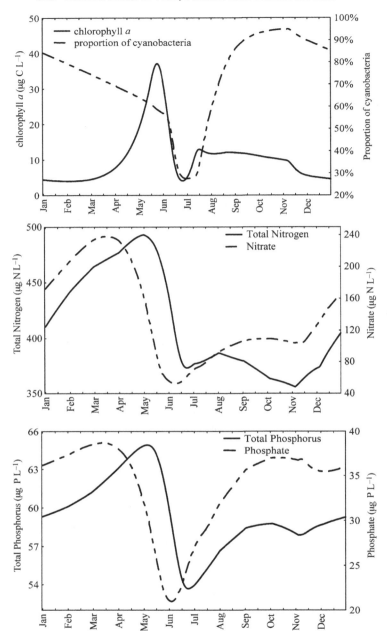

**Figure 4** Simulated epilimnetic chlorophyll, proportion of cyanobacteria, $PO_4^{3-}$, TP, $NO_3^-$, and total nitrogen (TN) annual cycles, based on nutrient loadings from the year 1962. Nutrient concentrations are averages weighted over the epilimnion and hypolimnion volume. Reproduced by permission of Elsevier from Arhonditsis and Brett (2005a, b). Copyright © 2005, Elsevier.

**Table 2** Monte Carlo analysis of the model based on current nutrient loading conditions and two scenarios of increased nutrient loading. Summary statistics of the annual values of the epilimnetic phytoplankton biomass ($\mu$g chl$a$ L$^{-1}$), proportion of cyanobacteria (%), total nitrogen ($\mu$g N L$^{-1}$), nitrate ($\mu$g N L$^{-1}$), total phosphorus ($\mu$g P L$^{-1}$), and phosphate ($\mu$g P L$^{-1}$).

| Condition | Statistic | Phytoplankton biomass | Proportion of cyanobacteria | Total nitrogen | Nitrate | Total phosphorus | Phosphate |
|---|---|---|---|---|---|---|---|
| Current conditions | Mean | 3.98 | 21 | 354 | 171 | 17.2 | 7.4 |
|  | Standard deviation | 0.26 | 2 | 53 | 32 | 1.9 | 1.4 |
|  | Range | 3.42–4.68 | 15–25 | 249–470 | 111–237 | 12.9–22.9 | 4.5–11.6 |
| Scenario 1 | Mean | 4.88 | 27 | 346 | 144 | 23.7 | 10.8 |
|  | Standard deviation | 0.2 | 2 | 54 | 31 | 2.8 | 2.1 |
|  | Range | 4.44–5.46 | 22–35 | 243–465 | 89–210 | 17.6–32.2 | 6.6–16.5 |
| Scenario 2 | Mean | 5.04 | 28 | 346 | 138 | 24.3 | 10.9 |
|  | Standard deviation | 0.26 | 3 | 53 | 27 | 2.9 | 2.1 |
|  | Range | 4.48–5.84 | 24–40 | 243–463 | 89–196 | 18.1–33.1 | 6.6–16.9 |

phytoplankton community and their proportion during the summer-stratified period exceeded 50%. There was also a decreasing trend for epilimnetic nitrogen, and especially $NO_3^-$ concentrations, which was about 10% lower than current conditions. In contrast, annual TP and $PO_4^{3-}$ concentrations increased on average by 6 and 3 µg P L$^{-1}$, but phosphorus was still the limiting nutrient of the system. Under the enrichment scenarios the system was characterised by larger spring phytoplankton blooms ($\approx$15 µg chl$a$ L$^{-1}$), and more frequent late summer to early autumn secondary blooms ($\approx$ 5–10 µg chl$a$ L$^{-1}$).

Interesting relationships were found between phytoplankton, TP, and $PO_4^{3-}$ during the summer-stratified period (Table 3). The positive correlation coefficients between phytoplankton biomass and TP ($>0.315$) indicate that due to the low summer phosphorus levels most of the available phosphorus is sequestered in phytoplankton cells. During the present phosphorus-limiting conditions, the rapid phosphorus uptake results in a negative correlation between algal biomass and $PO_4^{3-}$ ($r = -0.267$, $p = 0.001$) which, however, weakens and becomes insignificant for the first ($r = 0.009$, $p = 0.916$) and second ($r = -0.069$, $p = 0.403$) nutrient loading scenarios. The relaxation of the phosphorus limitation during the summer-stratified period was most evident between cyanobacteria biomass, TP, and $PO_4^{3-}$. Being the weakest phosphorus competitors, cyanobacteria benefit less from the available phosphorus in the water column, and under the current conditions their correlation is non-significant ($r = -0.106$, $p = 0.196$) with TP and significantly negative ($r = -0.551$, $p < 0.001$) with $PO_4^{3-}$. However, under the two enrichment scenarios, the phosphorus supplies to the Lake Washington epilimnion relax the cyanobacteria competitive handicap, and therefore the correlation of their biomass with TP increased significantly for both the first ($r = 0.170$, $p = 0.037$) and second ($r = 0.323$, $p < 0.001$) scenarios.


## Heuristic value of the model

One of the assumptions of the model is that zooplankton has the ability to maintain its somatic elemental (C:N:P) ratios constant by independently adjusting the production efficiency (biomass produced per food ingested) for carbon, nitrogen, and phosphorus (Elser and Urabe, 1999). Based on this concept, the stoichiometric theory predicts that zooplankton with low body C:P and N:P ratios recycle nutrients at higher C:P and N:P ratios than zooplankton taxa with high somatic C:P and N:P ratios (an explanation of this prediction is provided in Figure 5).

Hence, the assumption of constant C:N:P stoichiometries for the two zooplankton groups resulted in an increased model sensitivity to the non-limiting elemental (carbon and nitrogen) recycling. For example, the calculated fluxes of egested unassimilated material from the calibrated model were 5692 and 1064 10$^3$ kg yr$^{-1}$ of excess carbon and nitrogen during zooplankton feeding and corresponded to 83%

**Table 3** Monte Carlo analysis of the model based on current nutrient loading conditions and two scenarios of increased nutrient loading. Correlation matrix between the summer epilimnetic phytoplankton/cyanobacteria biomass and water temperature, total nitrogen, nitrate, total phosphorus, phosphate, total zooplankton, cladocerans and copepods biomass

| Condition | Organism | Water temperature | Total nitrogen | Nitrate | Total phosphorus | Phosphate | Zooplankton | Cladocerans | Copepods |
|---|---|---|---|---|---|---|---|---|---|
| Current conditions | Phytoplankton | −0.273 (0.001) | 0.526 (<0.001) | 0.392 (<0.001) | 0.315 (<0.001) | −0.267 (0.001) | 0.176 (0.031) | −0.512 (0.001) | 0.480 (<0.001) |
| | Cyanobacteria | −0.774 (<0.001) | 0.854 (<0.001) | 0.807 (<0.001) | −0.106 (0.196) | −0.551 (<0.001) | 0.750 (<0.001) | 0.000 (0.996) | 0.885 (<0.001) |
| Scenario 1 | Phytoplankton | 0.174 (0.033) | 0.070 (0.396) | −0.082 (0.321) | 0.494 (<0.001) | 0.009 (0.916) | −0.255 (0.002) | −0.751 (<0.001) | 0.088 (0.286) |
| | Cyanobacteria | −0.273 (0.001) | 0.416 (<0.001) | 0.293 (<0.001) | 0.170 (0.037) | −0.273 (0.001) | 0.153 (0.061) | −0.693 (<0.001) | 0.482 (<0.001) |
| Scenario 2 | Phytoplankton | 0.040 (0.624) | 0.187 (0.022) | −0.002 (0.976) | 0.443 (<0.001) | −0.069 (0.403) | −0.084 (0.305) | −0.678 (<0.001) | 0.244 (0.003) |
| | Cyanobacteria | −0.231 (<0.001) | 0.393 (<0.001) | 0.225 (0.006) | 0.323 (<0.001) | −0.131 (0.110) | 0.163 (0.047) | −0.569 (<0.001) | 0.457 (<0.001) |

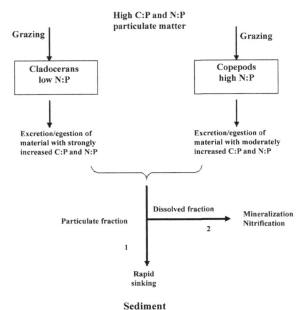

**Figure 5** Conceptual model of the zooplankton elemental composition effects on the N: P ratio of the excreted/egested material. During calibration, the assignment of higher values to the dissolved nitrogen fraction (arrow 2), increased greatly the ammonium levels and enhanced particularly the summer hypolimnetic accumulation. These results reflect the stoichiometric theory prediction that if homeostasis is maintained via post-assimilation processing and differential excretion of nutrients in dissolved form, then zooplankton recycling will tend to increase the retention time of the non-limiting elements in the water column. In this study, the most effective calibration strategy was to apportion the majority of the recycled nitrogen to the particulate fraction (arrow 1). This partitioning corresponds to the hypothesis that zooplankton homeostasis is also maintained during the digestion and assimilation process, i.e. by removing elements in closer proportion to zooplankton body ratios than to the elemental ratio of the food. Reproduced by permission of Elsevier from Arhonditsis and Brett (2005a, b). Copyright © 2005, Elsevier.

[egestion: (egestion + basal metabolism)] of zooplankton recycling in the system (Arhonditsis and Brett, 2005b). The partitioning of this material into the particulate/dissolved and inorganic/organic pools allowed: (a) examination of the potential consequences of different predictions of the stoichiometric theory, and (b) to offer a plausible explanation of the increasing alkalinity trends in Lake Washington.

(a) The assignment of higher values to the dissolved nitrogen fraction (DON and $NH_4^+$) increased ammonium concentrations to much higher levels than those observed in the system and particularly enhanced summer hypolimnetic accumulation. These results reflect the stoichiometric theory prediction that if homeostasis is maintained via post-assimilation processing and differential excretion of nutrients in dissolved form, then zooplankton recycling will tend to increase the retention time of the non-limiting elements in the water column (Elser and Foster, 1998). During the calibration, several approaches were followed to solve this problem, i.e. increase of the ammonium inhibition to nitrate uptake, lower mineralisation rates and higher nitrification rates, and none of these approaches proved to be more effective than simply assuming a higher particulate fraction of the egested material. This partitioning renders support to the hypothesis that zooplankton homeostasis is maintained during the digestion and assimilation process, i.e. by removing elements in closer proportion to zooplankton body ratios than to the elemental ratio of the food (Elser and Foster, 1998). A particulate fraction of 65% resulted in sedimentation rates that were within the observed ranges from past studies in Lake Washington (Edmondson and Lehman, 1981). However, the fact that these observations only overlap partially with the period when *Daphnia* became a prominent member of the zooplankton community indicates that further research is required to verify these rates.

(b) Increased nitrogen sedimentation and accumulation to the lake bottom might also have changed the relative magnitudes of the various sediment (nitrification, denitrification) processes. For example, model results showed that increased nitrification rates would have improved the agreement between simulated and observed nitrate hypolimnetic concentrations (Table 1). If the simulated nitrate deficit is actually a result of increased nitrification (sediments or water column), which is unaccounted for by the model, this needs to be supported by field data. Moreover, greater particulate deposition rates might have promoted sediment denitrification, which in turn can – at least in part – explain the significant alkalinity increase in the lake (Edmondson, 1994, figure 2E). This explanation deviates from Edmondson's (1994) hypothesis that the alkalinity increases were most likely induced by land-cover changes in the watershed, which altered chemical inputs to streams. Whether the relative magnitudes of nitrification/ denitrification have changed in the lake after the resurgence of P-rich *Daphnia* and their net effect (decrease of alkalinity by nitrification and increase by denitrification) can be associated with the observed alkalinity patterns, are issues that can be investigated via the current monitoring of the lake.

# References

Arhonditsis, G.B. and Brett, M.T. 2005a. Eutrophication model for Lake Washington (USA): Part I – model description and sensitivity analysis. *Ecol. Model.*, **187**, 140–178.

Arhonditsis, G.B. and Brett, M.T. 2005b. Eutrophication model for Lake Washington (USA): Part II – Model calibration and system dynamics analysis. *Ecol. Model.*, **187**, 179–200.

Arhonditsis, G., Brett, M.T. and Frodge, J. 2003. Environmental control and limnological impacts of a large recurrent spring bloom in Lake Washington, USA. *Environ. Manage.*, **31**, 603–618.

Edmondson, W.T. 1994. Sixty years of Lake Washington, a *curriculum vitae*. *Lake Reserv. Manage.*, **10**, 75–84.

Edmondson, W.T. 1997. Aphanizomenon in Lake Washington. *Arch. Hydrobiol. Suppl.*, **107**, 409–446.

Edmondson, W.T. and Lehman, J.T. 1981. The effect of changes in the nutrient income on the condition of Lake Washington. *Limnol. Oceanogr.*, **26**, 1–29.

Elser, J.J. and Foster, D.K. 1998. N, P stoichiometry of sedimentation in lakes of the Canadian shield: relationships with seston and zooplankton elemental composition. *Ecoscience*, **5**, 56–63.

Elser, J.J. and Urabe, J. 1999. The stoichiometry of consumer-driven nutrient recycling: theory, observations, and consequences. *Ecology*, **80**, 735–751.

Rykiel, E.J. 1996. Testing ecological models: the meaning of validation. *Ecol. Model.*, **90**, 229–244.

# References

Arhonditsis, G.B. and Brett, M.T. 2004. Evaluation of the current state of mechanistic aquatic biogeochemical modelling. *Mar. Ecol-Prog. Ser.*, **271**, 13–26.

Beck, M.B. 1987. Water-quality modeling – A review of the analysis of uncertainty. *Water Resour. Res.*, **23**, 1393–1442.

Bellinger D., Leviton, A., Waternaux, C., Needleman, H. and Rabinowitz, M. 1988. Low-level lead exposure, social class, and infant development. *Neurotox. Teratol.*, **10**, 497–503.

Bernard, S.M. and McGeehin, M.A. 2003. Prevalence of blood lead levels > or = 5 micro g/dL among U.S. children 1 to 5 years of age and socioeconomic and demographic factors associated with blood of lead levels 5 to 10 micro g/dL, Third National Health and Nutrition Examination Survey, 1988–1994. *Pediatrics*, **112**, 1308–1313.

Beven, K. and Binley, A. 1992. The future of distributed models – model calibration and uncertainty prediction. *Hydrol. Process.*, **6**, 279–298.

Bhavsar S.P., Diamond M.L., Gandhi N., and Nilsen J. 2004. Dynamic coupled metal TRANSport–SPECiation (TRANSPEC) model: application to assess a zinc contaminated lake. *Environ. Toxicol. Chem.* **23** (10), 2410–2420.

Chapra, S.C. 1997. *Surface Water-Quality Modeling*, McGraw-Hill, New York.

Chapra, S.C. and Canale, R.P. 1998. *Numerical Methods for Engineers*, 3rd edn, McGraw-Hill, New York.

Costanza R. and Sklar, F.H. 1985. Articulation, accuracy and effectiveness of mathematical models—a review of freshwater wetland applications. *Ecol. Model.*, **27**, 45–68.

Daniels, P.L. and Moore, S. 2002. Approaches for quantifying the metabolism of physical economies. *J. Industr. Ecol.*, **5**, 69–93.

Dillon, P.J. and Rigler, F.H. 1974. A test of a simple nutrient budget model predicting the phosphorus concentration in lake water. *J. Fish. Res. Board. Can.*, **31**, 1771–1778.

Fava, J., Consoli, F., Denison, R., Dickson, K., Mohin, T. and Vigon, B. (editors) 1993. *A Conceptual Framework for Life-cycle Impact Assessment*, SETAC and SETAC Foundation for Environmental Education Inc., Pensacola, FL, 11–26.

Franks, P.J.S. 1995. Coupled physical–biological models in oceanography. *Rev. Geophys.*, **33**, 1177–1187.

Gewurtz, S.B., Gandhi N., Stern G.A., Franzin W.G., Rosenberg B. and Diamond M.L. 2007. Dynamics of PCBs in the food web of Lake Winnipeg. *J. Great Lakes Res.*, **32** (4), 712–-727.

Hartmann, S. 1996. The world as a process: simulations in the natural and social sciences. In: *Simulation and Modelling in the Social Sciences from the Philosophy of Science Point of View*. R. Hegselmann, *et al.* (editors), Theory and Decision Library, Kluwer, Dordrecht, 77–100.

Hertzman, C. and Weins, M. 1996. Child development and long-term outcomes: a population health perspective and summary of successful interventions. *Soc. Sci. Med.*, **43**, 1083–1095.

Hewett, S.W. and Johnson, B.L. 1992. *A Generalized Bioenergetics Model of Fish Growth for Microcomputers*, University of Wisconsin Sea Grant Institute, Madison, WI.

Hornberger, G.M. and Spear, R.C. 1981. An approach to the preliminary analysis of environmental systems. *J. Environ. Manage.*, **12**, 7–18.

Jorgensen, S.E. and Bendoricchio. G. 2001. *Fundamentals of Ecological Modelling*, 3rd edn, Elsevier, Amsterdam.

Jorgensen, S.E., Nielsen, S.N. and Jorgensen, L.A. 1991. *Handbook of Ecological Parameters and Ecotoxicology*. Pergamon Press, Amsterdam.

Leggett, R. W. and Williams, L. R. 1981. A reliability index for models. *Ecol. Model.*, **13**, 303–312.

Mackay, D., 2001. *Multimedia Environmental Models: the Fugacity Approach*, 2nd edn, Lewis Publishers, Boca Raton, FL.

Mayer, D.G. and Butler, D.G. 1993. Statistical validation. *Ecol. Model.*, **68**, 21–32.

Oreskes, N., Shrader-Frechette, K. and Belitz, K. 1994. Verification, validation, and confirmation of numerical models in the earth sciences. *Science*, **263**: 641–646.

Power, M. 1993. The predictive validation of ecological and environmental models. *Ecol. Model.*, **68**, 33–50.

Reckhow, K.H., Clements, J. T. and Dodd, R.C. 1990. Statistical evaluation of mechanistic water-quality models. *J. Environ. Eng.*, **116**, 250–268.

Vollenweider, R.A. 1975. Input–output models with special reference to the phosphorus loading concept in limnology. *Schweiz. Z. Hydrol.*, **37**, 53–84.

# 7

# Presenting your work

## 7.1 Getting started – strategies for successful writing

The most frequently heard gripe concerning writing is 'I just cannot get started'. There is unfortunately, no 'magic formula' to resolve this perennial problem, and different strategies work for different people. The approach that one of the authors of this text uses to write (including this book), is first to write out an outline plan of what he wishes to say, before starting to write on any aspect. An important factor is that he does not worry about grammar, dictionaries or thesauruses (or is that thesauri?), or even necessarily that what is being written actually makes much sense, the important thing is to get some of his ideas down on paper (or more correctly disk). At the end of the writing session, the product is printed out and reviewed. While by no means all of what is written may survive into the finished version, this method does give the satisfaction (however illusory it may sometimes be) of having made progress. Undoubtedly, the advent of word-processing software has facilitated greatly this approach, allowing as it does, the author to write chunks of prose in no recognisably logical order, only to subsequently 'cut and paste' into a coherent narrative.

In short, a massive task will always seem massive if you never start to tackle it. By writing anything, you make inroads into the task, and gradually make it more manageable. Of all the possible strategies for writing, procrastination is the least successful.

## 7.2 How to write your dissertation

After a long gestation period, you will be required to deliver a healthy, bouncing 2 kg dissertation. The analogy to giving birth is arguably apposite, given the myths that surround both processes, and the anxiety that both induce.

*Student Projects in Environmental Science*   Edited by Stuart Harrad and Lesley Batty
© 2008 John Wiley & Sons Ltd

Most students feel overwhelmed by the magnitude of the task facing them, with one of the key factors being their lack of experience. By definition, undergraduate students will never (with a very few exceptions) have written a dissertation before. It therefore makes sense to take every opportunity to practise. In this respect, many universities require undergraduate students to submit (sometimes formally assessed) literature reviews or research proposals before commencing their research project. Besides providing you with formal feedback on your progress, such reviews/proposals offer an ideal opportunity to practise for the 'real thing'. After all, in essence, the dissertation is just another piece of coursework – albeit a very important one. If written with care, these pieces of work can form a useful starting point for your dissertation.

Another useful exercise is to obtain copies of past (successful) dissertations on your course or in your subject area. Your supervisor or course tutor should be able to help you with this. One of our own courses includes such an exercise as part of a research methodology module for students about to commence their research projects. The purpose is to for students to critically evaluate previous dissertations (both good and bad), in an effort to enhance their awareness of what it is that makes a dissertation good, bad, or indifferent. Even if you are not required to conduct such an exercise as part of your degree programme, then we recommend that you undertake such an evaluation yourself, and discuss your opinions with your supervisor.

One obvious but all too frequently overlooked source of guidance as to the layout and format of your dissertation are the instructions provided as part of your course. Whatever it is called (e.g. 'Final Year Research Project Manual') it will almost certainly contain details of how many words it should contain, as well as presentational issues such as line spacing, font size and type, margins etc. Given the weighting assigned to such issues in many marking schemes, you would be foolish to ignore such guidelines. After all, it does not take a genius to adhere to instructions of this nature, and doing so should prevent needless loss of marks.

When writing your dissertation, it is useful to visualise the process as that of telling a story, the basic narrative of which may be summarised thus:

- *What are you trying to do?* A common problem is that students do not state clearly what their objectives are. This stems frequently from the fact that the student believes them to be 'obvious'. This is a big mistake. It may be 'obvious' to the student, but will not be necessarily so to the examiner(s). Certainly, the examiner(s) can elucidate the objectives with detailed reading of the dissertation, but this is hardly calculated to make them happy and inclined to award high marks. On a more fundamental level, a clear concise statement of objectives is a great help in writing a dissertation, as it helps keep the author focused. A 'top tip' is thus to write and print out a page stating your objectives, and to stick it on the wall next to your PC.

- *Why are you trying to do it?* Addressing this question involves describing why your research project is justified. Aspects to consider could include one and usually more than one of the following: human or ecotoxicity of a pollutant; the existence of inconsistencies in previous work that require resolution; a lack of knowledge about the sources of a pollutant; an observation of an unexplained decline in population of a particular species etc.

- *How did you do it?* Quite simply, what methods did you use to conduct your research. This will include detailed but concise descriptions of sampling and measurement methodology. Where appropriate it will also incorporate details of statistical analyses, or questionnaire format etc.

- *What did you find out?* This involves summarising your data. The summarising aspect is especially relevant where large datasets are produced. In such instances, it is common practice to include comprehensive tabulations of data as Appendices, with the option of providing full datasets in electronic formats on a CD becoming an increasingly feasible (and desirable) alternative.

- *What does it mean?* This is probably the hardest, yet the most rewarding aspect of producing a dissertation. It involves making sense of your hard-generated data. This is a point where reference to your objectives can be of great help in focusing your efforts. How does your data help to answer the questions posed? Importantly, whilst addressing your objectives should provide the basic framework for your data interpretation, you should not close your mind to the possibility of previously unforeseen ideas and concepts emerging from your data.

In conventional scientific reporting format, points 1 and 2 would typically form the introductory chapter; point 3 the methods or experimental chapter; and points 4 and 5 the results and discussion chapter(s).

One of the best reasons for writing your dissertation 'as you go', is that like planning (see section 1.5), it forces you to think about your work. In order to present your work in a clear and comprehensible fashion, you have no alternative but to understand it. In this respect, it is analogous to teaching; in order to explain a topic well to others, you have to understand it thoroughly.

## 7.2.1 Choosing a title

The trick here is to choose an informative but concise title. While you should avoid impossibly vague titles such as 'A Study of Global Warming', you should similarly steer clear of long-winded titles like 'An Investigation into the Relationships between Global Warming, Ocean Temperature, Emissions of Dimethylsulfide by Marine Phytoplankton, Formation of Cloud Condensation Nuclei, and Cloud Albedo'. A more concise version of this could read 'Marine Phytoplankton, Dimethylsulfide, Cloud Albedo and Climate Change'. Discuss your title with your supervisor.

## 7.2.2 Writing an abstract

An important aspect of your dissertation is the abstract. Its length will vary between courses (typically 150–250 words), but its fundamental purpose and thus its content remains constant. All too often, the abstract is thought of as something of an afterthought. This is a mistake, although clearly it cannot be written until you have completed the bulk of your dissertation, and have arrived at your conclusions. The importance of your abstract is that it represents the part of your dissertation that people will read first. If written well, it will encourage them to read further. If not, then your abstract will represent the only part of your dissertation that is read by anyone except you, your supervisor, any other markers, and – if they are particularly proud (not to mention masochistic) – your parents/partner.

The abstract is intended to summarise the dissertation. Its basic structure must therefore match that of the dissertation as a whole (see above), but in a significantly pared down form. A good way of visualising this paring down process is that every chapter in your dissertation, should be summarised as a paragraph in your abstract. Typically therefore, your abstract should comprise the following paragraphs describing:

- what your research question/hypothesis is, and (very briefly) why it needs to be addressed

- what methods you used to answer the question/test the hypothesis

- what your results are

- what your results mean – in particular, how they relate to your original question/ hypothesis.

The following example, offers an unfortunately excellent example of how NOT to write your abstract.

'The purpose of this investigation is to investigate how much heavy metals people were exposed to via breathing indoor air. Results are presented and conclusions drawn'.

While this has the benefit of being concise, it fails summarily to address any of the points A-D identified in the above checklist. The author failed to make clear the hypothesis actually being tested, *viz* that human inhalation exposure to cadmium, lead and mercury (note that the specific heavy metals studied were not mentioned) is greater indoors than outside. They omitted to mention the health concerns that motivate studies of human exposure to these elements, overlook completely any mention of how exposure was measured, and the size and nature of the population group studied. They then commit the cardinal sin of failing to state anything about what they found out, and therefore give no indication of the conclusions of their study.

Overall, readers of this abstract – and remember that this is almost certainly the FIRST thing that the examiners will read – will gain very little idea of what the dissertation contains, and the complete lack of any results or conclusions raises suspicions that nothing has been achieved. Given that 1st impressions are very difficult to shake off, you would be wise to ensure your abstract gives the best possible summary of your work.

The hypothetical example below illustrates what the abstract should have read like.

'Elevated human exposure to heavy metals like cadmium, lead, and mercury is considered a potential health risk. Given the high proportion of time spent by UK adults indoors, this study evaluated the relative significance of exposure to Cd, Pb, and Hg via inhaling indoor compared to outdoor air. To do this, airborne concentrations of these elements were determined in a number of representative indoor environments (homes, offices etc.). These were combined with previously published data on concentrations in outdoor air, and information on the relative proportions of time spent by typical UK adults outdoors and in different indoor environments. The results show that inhaling indoor air contributes respectively 82, 67, and 83% of total inhalation exposure to Cd, Pb, and Hg respectively. As previous work has shown a typical UK adult to spend 93% of their time indoors, it is concluded that it is this excess of time spent indoors that makes inhalation of indoor air so important, rather than the existence of higher concentrations in indoor air.'

## 7.3 How to represent graphically your data

You will be aware of the well-known saying 'a picture paints a thousand words'. While this has some truth to it, many students make the mistake of including too many unnecessary graphs or Figures in their dissertation. You should not include graphs for the sake of it. Consider very seriously whether a Table would be better in some instances. A good illustration of when graphs are overused frequently is when one is examining whether there exists a linear relationship between an environmental parameter (say temperature) and the concentrations of a number (e.g. 20) pollutants. Students often make the mistake of including 20 separate scattergraphs plotting concentration of each individual pollutant versus temperature – a single illustration is given as Figure 7.3.1. This takes up a large amount of space in the dissertation and makes life very difficult for the reader (the examiner!). In such instances, you should include one such graph as an illustrative example, but then summarise the salient data obtained from the graph (i.e. the slope, the intercept, the correlation coefficient, and most importantly, the p value, which indicates the strength of correlation between concentration and temperature – see section 5.1) in a single Table. An example is given as Table 5.1.3.

Sometimes, you have choices as to what sort of graph to use to illustrate the patterns in your data. For example, Figure 7.3.2 is an alternative way of plotting the same dataset plotted in the scattergraph that is Figure 7.3.1. Which of the two styles most clearly demonstrate to you that there is a statistically significant relationship between temperature and pollutant concentration? There is no right answer to this

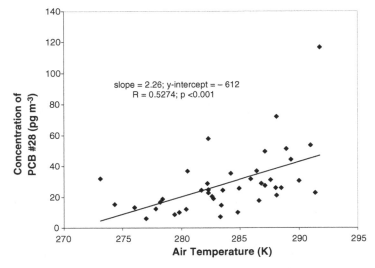

**Figure 7.3.1** Scattergram representing the relationship between concentration of PCB #28 and temperature.

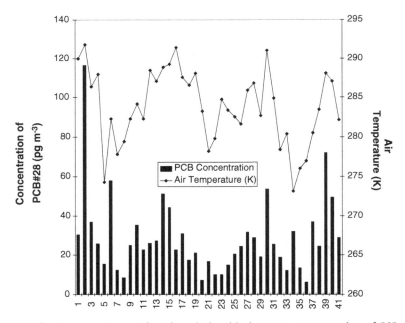

**Figure 7.3.2** Scattergram representing the relationship between concentration of PCB #28 and temperature.

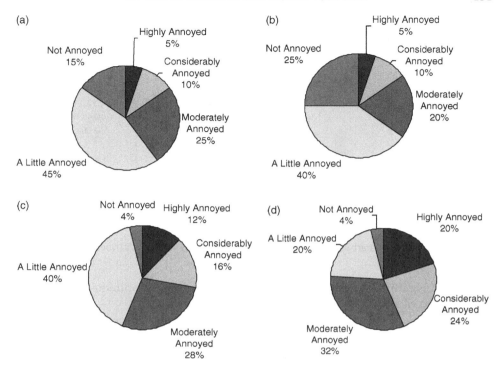

**Figure 7.3.3** Pie chart representation of annoyance levels amongst age group of: (a) < 30 Years; (b) 31–45 years; (c) 46–60 years; and (d) > 60 years.

question, as people's perceptions vary, but for the record, the author of this chapter prefers the scattergraph (Figure 7.3.1).

Figures 7.3.3a–d, 7.3.4, 7.3.5 and 7.3.6 show how the same data can be illustrated in different ways. In this instance, the data used records the percentage of people living in the vicinity of an airport that report one of the following categories of reaction to aircraft noise: highly annoyed, considerably annoyed, moderately annoyed, a little annoyed, or not at all annoyed (see also section 5.1). The hypothesis to be tested is that older people are more annoyed, for instance because they spend more time at home and have greater exposure to the noise than younger people who are out at work. To test the hypothesis, the annoyance category data was split into response from 4 age groups: < 30, 30–45, 45–60, and 60+. Figures 7.3a–d are 4 pie charts, each showing the responses of each age group.

By comparison, Figure 7.3.4 is a single bar chart, showing the same data, while Figure 7.3.5 is a stacked bar chart that shows how the proportion of responses within an annoyance level category depends on age group. Which do you think most clearly represents the fact that the 60+ age group does appear to be more annoyed than the other age groups? Again, there is no right answer. Figure 7.3.6 combines both graphical and tabular information. Specifically, it supplements a single bar chart

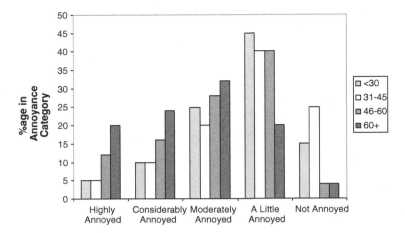

**Figure 7.3.4**  Bar chart representation of annoyance levels amongst various age groups.

like that in Figure 7.3.4, with a table of the data plotted. In the opinion of the author, this is an excellent way of illustrating the point to be made.

Each of these graphs was produced in Excel. As mentioned elsewhere, for detailed guidance as to produce graphs in Excel, you are referred to the text of Keller (2001). A useful exercise is to take the data given in the Table that is in Figure 7.3.6 and use Excel to reproduce Figures 7.3.3a–d, 7.3.4, 7.3.5, and even 7.3.6. Spend some time experimenting with different ways of representing graphically these data, and decide which portrays the 'message' of the data most clearly. This approach is one you should follow when deciding how to present your own data. If you are unsure as to which type of graph to use, then ask some friends for their opinion.

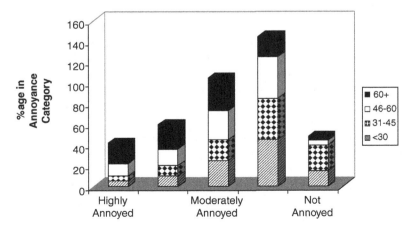

**Figure 7.3.5**  Stacked bar chart representation of the percentage of various age groups falling into various annoyance categories.

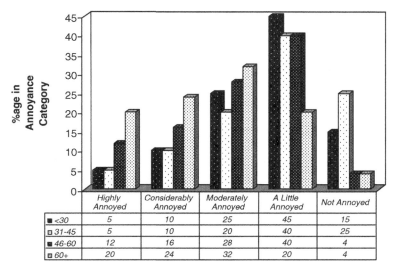

**Figure 7.3.6** Bar chart representation of annoyance levels amongst various age groups with data table.

## 7.4 How to cite references

There are several widely used referencing systems available to you. While some courses will insist on a particular scheme, others may not. In such instances, whichever system you choose, you must apply it consistently throughout your dissertation.

### 7.4.1 How to reference within your text

The following paragraph illustrates how to make reference to other relevant work within the text of a document.

'In line with many other regulatory agencies world wide, the UK's Food Standards Agency (2000) considers that consumption of fish constitutes an important pathway of human exposure to polychlorinated biphenyls (PCBs). Consequently, there have been numerous attempts to predict their transfer through freshwater aquatic food chains (e.g. Abbott *et al*, 1995; Harrad and Smith, 1998; Thomann, 1989). Recently, there have been significant advances in the predictive power of such mathematical models (Thomann, 1999a,b), which have been reviewed comprehensively by Gobas (2001, pp. 145–163).'

Where you cite more than one reference, note that each must be separated by a semi-colon (as at the end of the second sentence). In the example here, the three references are cited in alphabetical order according to first author's name; equally correct is placing them in chronological order according to publication date. Whichever you choose, you must be consistent throughout. Note also that where there are two co-authors of a paper, both names are cited, but when the number of co-authors exceeds two, then only the first author is cited with the existence of co-authors denoted by *et al.* The final sentence illustrates how to cite two or more papers published in the same year by the same author(s), as well as how to refer to a particular section of a publication – in this case a chapter of a book.

## 7.4.2 How to prepare your references list

In the course of your dissertation you will probably cite a wide variety of categories of publications. The Harvard format for the main categories of publication are shown below. Again, slight variations on these formats exist (e.g. sometimes the date of publication is parenthesised, or the authors' initials are placed before the surname). However, unless your course requires a specific system to be employed, it is perfectly acceptable to use such alternatives, but remember that consistency is paramount. As will be seen, the protocols for referencing some of the more esoteric categories of publication are less strict. In such cases, remember that the purpose of a reference list is to enable the reader to locate the publication cited, and therefore, when in any doubt as to how to reference, record as much information as possible.

Articles in research journals should be listed thus:

'Author(s) – initials – date of publication – title of paper – journal name – volume – start and end page numbers. For example: Harrad, S. and Smith, D. 1998. Field evaluation of a mathematical model of PCB transfer through the freshwater aquatic food chain. *Sci. Tot. Environ.*, **212**, 137–144.'

Note that the journal name is italicised (or underlined or emboldened). It can be abbreviated if lengthy, but the abbreviation should be the official one used by the journal and which will be present on the paper cited. For example, in the illustrative reference above, the journal cited is *The Science of the Total Environment*. If any doubt exists as to the correct abbreviation, then cite the full name. Note the need for a full stop after each abbreviation of a word to denote that, e.g. *Sci.* is an abbreviation of Science.

References to whole non-edited books, i.e. where – like this one – the entire book has a single author or authors, should be made thus:

'Author(s) – initials – date of publication – title of book – edition number if appropriate – publishers and place of publication. For example: Mackay, D. 1991. *Multimedia Environmental Models The Fugacity Approach*, Lewis, Chelsea, MI.'

The same system is used (but making reference to the editor) for edited books (these usually consist of a number of chapters based on a common theme but each written by different authors):

'Harrad, S. (editor) 2001. *Persistent Organic Pollutants: Environmental Behaviour and Pathways of Human Exposure*, Kluwer Academic, Norwell, MA. ISBN 0-7923-7227-1'

Where one wishes to cite an individual chapter within an edited book, the style to be employed is:

'Chapter author(s) – initials – date of publication – chapter title – book editor(s) – initials – book title and edition number if appropriate – publishers and place of publication – start and end page numbers of chapter. For example: Harrad, S. 2001. The environmental behaviour of persistent organic pollutants. In: *Pollution: Causes, Effects and Control*, 4th edn, Harrison, R. M. (editor), Royal Society of Chemistry, Cambridge, 445–473.'

### 7.4.3   What to include in your references list and how to present it

Very simply, anything that you have cited in your text must be included in your references list. Where you have read a publication (e.g. a textbook in order to get 'background' information on your dissertation topic), but have not made direct reference to it in the text, then you may denote this optionally by including references to such publications in a *Bibliography*.

An example reference list is given below. Note that it also includes references to two reports by government agencies (the UK HMIP and the US EPA respectively). The protocols for referencing such publications are flexible, and in these cases, the maximum information has been cited. No ISBN numbers were assigned to these reports, but were they available, these would have been included, as they provide unique 'tracers' to facilitate location of the publication.

## References

Abbott, J.D., Hinton, S.W. and Borton, D.L. 1995. Pilot scale validation of the river/fish bioaccumulation modeling program for nonpolar hydrophobic compounds using the model compounds 2,3,7,8-TCDD and 2,3,7,8-TCDF. *Environ. Toxicol. Chem.*, **14**, 1999–2012.

Ayris, S., Currado, G., Smith, D. and Harrad, S. 1997. GC/MS procedures for the determination of PCBs in environmental matrices. *Chemosphere*, **35**, 905–917.

Craig, J.F. 1993. *Pike – Biology and Exploitation*, Chapman and Hall, London.

Hawker, D.W. and Connell, D.W. 1988. Octanol–water partition coefficients of polychlorinated biphenyl congeners. *Environ. Sci. Technol.*, **22**, 382–387.

HMIP. 1996. *Risk Assessment of Dioxin Releases from MWI Processes*. Report HMIP/CPR2/41/1/181, Her Majesty's Inspectorate of Pollution.

Murphy, T.J. and Sweet, C.W. 1994. Contamination of Teflon surfaces in the atmosphere. *Atmos. Environ.*, **28**, 361–364.

Swackhamer, D.L. and Eisenreich, S.J. 1991. Processing of organic contaminants in lakes. In: *Organic Contaminants in the Environment*, Jones, K.C. (editor), Elsevier, Oxford, 33–86.

Thomann, R.V. 1989. Bioaccumulation model of organic chemical distribution in aquatic food chains. *Environ. Sci. Technol.*, **23**, 699–707.

Thomann, R.V. and Connolly, J.P. 1984. Model of PCB in the Lake Michigan lake trout food chain. *Environ. Sci. Technol.*, **18**, 65–71.

US EPA. 1994. *Estimating Exposure to Dioxin-Like Compounds, Volume III – Site Specific Procedures*. Report No. EPA/600/6/-88/005Cc (External Review Draft), US Environmental Protection Agency, Washington, DC.

### 7.4.4   How to include internet publications in a reference list

It is imperative that as well as the URL (i.e. the web address – e.g. http://www.bham.ac.uk), you cite the date of accessing a web-based source of information. This is because they are updated/removed regularly. A key principle to remember always with websites is that the information on them is not necessarily peer-reviewed, and may therefore not be wholly valid scientifically. Government websites are usually authoritative sources of information, but relying solely on websites for information is most definitely not recommended, and you should always use textbooks and journals as well.

## 7.5   How to defend your work in an oral exam

Some courses require oral or *viva voce* exams for undergraduate research project students. Although such oral exams will only form a relatively minor percentage of your overall mark, and their duration (typically 15–30 minutes) appreciably shorter than the PhD equivalent, the principles of how to prepare for and behave in a viva apply equally at all levels. As the format of undergraduate level vivas is less uniform between institutions and even courses, we have not attempted to cover the administrative details. Instead, we recommend that you ask your supervisor and/or course tutor for advice as to what will happen and what you need to do.

An important thought to bear in mind is that for undergraduate research projects, only a relatively small proportion (e.g. 20%) is awarded on the basis of the viva. This does not mean that you should not try to perform to the best of your ability, but that should you for whatever reason – e.g. nerves – perform below par, the consequences are unlikely to be catastrophic. The purpose of this section is thus to demystify the process and suggest approaches to clearing this last hurdle.

Our first approach is to consider an example of what not to do. While it is perfectly reasonable (and indeed expected) of the candidate to defend their work from undue criticism, we would not recommend the over robust approach taken by one student in his PhD exam. In this instance, the candidate took such exception to the criticism of one examiner, that the student had to be physically restrained from attacking the examiner! Whether or not the criticism was justified, it is clear (hopefully) that such behaviour was unlikely to dissuade the examiner[1]. In short, a golden rule is to keep calm.

A key to maximising your performance in the viva is an understanding of what happens. After submission of your dissertation for examination, a copy is sent to all those involved in the viva. The exact composition of the examining 'team' varies from university to university, but usually consists of your supervisor and one other member of staff. You will be notified in advance of the date, time and venue of the

---

[1]Readers may be glad to learn that following revision of the controversial section, resubmission, and a 2nd (thankfully non-combative) viva, the student passed.

viva. The examiners will then read the dissertation in depth, and formulate questions and issues for discussion. Clearly, there is no formulaic set of questions, with each project and examiner having their own individual perspective. However, students are recommended to put their dissertation to one side for a couple of weeks after submission, before re-reading it with the express purpose of analysing its key points and arguments. This approach should help the student identify any issues that are likely to provoke discussion. These are not necessarily 'weak' points, but the novel aspects of the work which the student must be prepared to defend. Having above ruled out the 'overly robust' approach to defending your work, it should be stressed that the student should politely but firmly defend their work where necessary. For example a reply along the lines of 'Yes, that is an explanation that I had considered. However, if you look at what I have written on Page xx, you will see that on close examination, the evidence of Table xx does not support it, because of. . . .'.

If your supervisor does not suggest it, you may wish to ask them to give you a 'mock' viva prior to the real event. The purpose of this will be to give you an idea of the kind of questions that may be asked, and to give you some experience of being 'locked in' a room and forced to talk about your work for 15–30 minutes.

### 7.5.1   What to wear?

While you are not being judged on your sartorial elegance, we recommend that you dress smartly. The vast majority of your examiners will be attired smartly, and even if they turn up in t-shirt, shorts, and sandals, you will not do yourself any harm by taking the sartorial high ground. Take the view that it is a special day for you, and dress accordingly.

### 7.5.2   Punctuality

Make sure that you are on time for your viva. While this is an obvious point, it is important to ensure that you have allowed sufficient time to make the journey from home to the viva venue, making allowance for unforeseen problems. The experience of one of the authors serves to illustrate the importance of this. His viva was set for 2 pm, so he left home at 1 pm for the 5 minute walk to the nearest bus stop to catch a bus at 1.10 pm. With buses running also at 1.25 and 1.40, and a journey time of 10 minutes, this seemed a very comfortable situation. Needless to say, when it got to 1.35 pm with no sighting of any bus, panic began to set in. A public phone box was about 100 metres away (mobile phones were not used widely at the time), and the author was weighing up his prospects of calling one of his colleagues to come and pick him up by car, against the possibility of a bus sailing past the bus stop while he made the call, when in the true tradition of British public transport, a small flotilla of buses arrived simultaneously. In the end, the author arrived with 5 minutes to spare, to be greeted by admiring comments from fellow students regarding his 'coolness'.

The moral of this story is that you should always allow twice as much time as think you will need, and if you have a car, check the petrol tank is full, and that you are a paid up member of the AA or RAC.

## 7.6  How to make effective oral presentations

Making a presentation is usually high on the list – if not top – of the activities most dreaded by students. While it is undoubtedly true that some people are more naturally gifted public speakers than others, it is equally true that by adhering to some basic principles, anyone can learn to make effective public presentations of their work. Whatever career path you follow after graduation, you will almost certainly require presentation skills. Indeed, many job selection processes involve some kind of oral presentation to the selection panel. To summarise, being able to make effective presentations is an extremely valuable transferable skill.

Whether you will have to make an oral presentation as part of the assessment process of your research project will depend on your specific course, but we have included this section as it is becoming increasingly common.

As with written modes of disseminating your work to a wider audience, a good starting point is to pitch it at an appropriate level for your audience. The amount of background material you would cover would be greater in a talk delivered to other students on your course (many of whom although scientifically 'literate', will not be familiar with your specific research topic), than would be necessary if your audience consisted of experts on your topic at an international conference. Another key factor shaping your presentation, will be the time available to you. At the risk of stating the obvious, use only the time allotted to you. A surprisingly frequently encountered problem with presentations is that the speaker tries to cram too much in to the talk. This results in overrunning the allotted time, and often results in the speaker having to skim over or even omit key aspects of the talk.

Just as with writing your dissertation, it is wise to sketch out a plan of your talk before you start to write. It should again follow the What, Why, How, What, What plot line format described in section 7.2.

The authors have been privileged to witness some greatly entertaining and effective public speakers, who have simply extemporised on their topic. For the rest of us mere mortals (the authors included), the use of visual aids, such as overhead projection apparatus, slide projectors, and video beams, is necessary.

It is highly probable that students will make use of commonly available presentation software such as PowerPoint. This allows the presentation to be broken down into a series of slides, each consisting of a heading, followed by a figure (graph, diagram or photograph), or text laid out as a series of bullet points. A common presentational fault is that of 'overcrowding' the slide. One way to avoid this is to abide by the maxim of 'present as you would like to be presented to'. do not put too much information on one slide. Not only will this overload your audience and lose their interest, but it can sometimes result in text that it is simply too small for your audience to read. To avoid this, a good strategy is to impose a physical restraint by keeping your font size to a minimum size of 24 point. Also, bear in mind that your bullet points do not have to be exact copies of what you will say. It is better to regard the bullet points as a kind of shorthand prompt to both the audience and importantly yourself.

Although there are no hard and fast rules as to the number of slides you should include, a good rule of thumb is that you should prepare no more than 1 slide for every minute of your talk. This is a maximum, and if your slide covers a particularly important aspect of your story, that requires detailed explanation, you should allow more time. Overall, one would do well to bear in mind the advice offered to one of the authors on starting out as a University lecturer, *viz* that however many slides you initially think you will need, you should halve this, in order to arrive at the optimum number. While not entirely serious, this maxim is surprisingly effective, because it makes allowance for the speaker needing more time to elaborate on more important points. This is something that inexperienced speakers find difficult to believe, given their initial view of the talk as a public ordeal that is best got over with as soon as possible. In reality, almost without exception, speakers find themselves caught up in the experience of having a captive audience for their ideas, and it is far more common for speakers to overrun than to finish earlier than scheduled.

When planning the timing of your talk, do remember to allow the correct amount of time for questions at the end. This will usually be specified in the invitation (or order) to speak. Facing questions at the end of a talk is held widely by many students to be a particularly cruel form of rubbing salt into a rather nasty wound. A basic rule to avoid disasters when fielding questions is not to attempt to answer questions that you do not understand, or do not know the answer to. This may seem obvious, after all, you would not answer a request for directions from a foreign visitor in a language you did not understand, nor would you give direction to a location you had never heard of. In the first instance, you would ask the questioner for clarification of the question in words that you did understand, and in the latter instance, you would apologise and state that unfortunately you did not know the answer. Fielding such questions at the end of a talk requires essentially the same approach, with the slight modification that one should only usually permit a questioner one attempt at rephrasing a question. If you were still unsure as to what you were being asked, it is perfectly acceptable to offer to discuss the question with your inquisitor in private after the talk.

In the majority of cases, questions are perfectly straightforward, and may for example, take the form of requests for more detail of your methods (exactly how many samples did you take?), or may be questions of the form 'while your suggested interpretation of your data may be correct, have you not considered the following alternative interpretation...'. Questions of the former type should be extremely straightforward, while your response to the latter category will either consist of: (a) an explanation that yes, you had considered the alternative, but discounted it for reasons x, y, and z, or (b) that you had not considered that particular alternative. If (b), you may – if you are sufficiently quick thinking – be able to explain why the proffered alternative is incorrect, or you could simply thank the questioner for an interesting suggestion, and suggest that you discuss it in more detail after the talk.

Once you are reasonably happy with your slides, find a quiet spot and run through your talk, timing yourself. By all means make notes that help prompt you, but by and large your slide bullet points should be sufficient for this purpose. As you practise, make notes of points where changes may be necessary. If you find you are finishing a few minutes too early, then resist the temptation to add more slides, and just try speaking more slowly. Once you have run through the talk several times and made necessary adjustments, then ask your supervisor or a colleague to view a dummy run. Ask what they think, and take on board their advice. This may all seem a lot of effort, but remember you are about to give a public performance, and no-one would expect an actor to perform without running through their script, familiarising themselves with the props, and extensive rehearsals.

Some miscellaneous dos and do nots

- do make discreet use of notes if you find that it helps

- do practise your talk, making sure you keep within the time limit

- do not speak to the screen, and do face the audience when you are speaking to them

- do not read a script, and certainly *do not* read from cards

- do not include too much text on one slide

- do not include too many slides

- do speak slowly and clearly

## 7.7  Summary

This chapter is designed to help you present the work that you have done for your dissertation in the best possible manner. If there is one maxim to which you should adhere when presenting your work, it is always to try and place yourself in the shoes of your audience (e.g. the examiners). By so doing, you should be able to ensure that you explain what you have found out, how you found it out and what it means in a clear, concise and easily comprehensible fashion, that will maximise your marks. Good luck!

# Index